GREEN WITNESS

GREEN WITNESS

ECOLOGY, ETHICS, AND
THE KINGDOM OF GOD

LAURA RUTH YORDY

 CASCADE *Books* · Eugene, Oregon

GREEN WITNESS
Ecology, Ethics, and the Kingdom of God

Cascade Books
A Division of Wipf and Stock Publishers
199 W. 8th Ave., Suite 3
Eugene, OR 97401

ISBN 13: 978-1-55635-335-2

Cataloging-in-Publication data:

Yordy, Laura Ruth.
        Green witness : ecology, ethics, and the kingdom of God / Laura Ruth Yordy.

        viii + 182 p.; 23 cm. Includes bibliographic refrences (p. 173–82).
        ISBN 13: 978-1-55635-335-2 (alk. paper)

        1. Environmental ethics. 2. Human ecology—Religious aspects—Christianity.
3. Environmentalism—Religious aspects. I. Title.

BT695.5 Y67 2008

Manufactured in the U.S.A.

CONTENTS

PREFACE

W HEN I was a bookish twelve-year-old (rather than a bookish middle-aged creature), my family moved to four acres of woods in Virginia. The new locale was not a rural area, but the acres of thick forest were a radical change from the tidy suburban development I had been accustomed to. I spent hours wandering the woods, exploring the creeks, trying to climb trees and walk on slippery logs. I learned to feel comfortable there, enjoying the seclusion (I thought—now I suspect my parents could always see me from the house) and the birdsongs. One afternoon, noticing a small hole in the streambank, I knelt and put my eye close to the mud to look inside. What I saw, to my great astonishment, was another eye, looking out at me. I was flabbergasted, and stood up so fast that I fell back into the water. (I don't know if seeing me had as strong an effect on the snake—probably not.) Without the words to describe it, I was faced (literally) with the realization that mine was not the only subjectivity in the woods. I was as much explored and observed as I was explorer and observer. The universe is not solely an object for humans to study, but a grand multiplicity of subjects, each perceiving and constructing its own understanding of the world.

Decades later, this experience—and my love of animals, plants, and wilderness—came into play as I discovered Christianity's rich visions of the relationship between Creator and creatures. But these visions did not seem visible to most church communities. For several years, I believed that education was the answer; that if faithful people were simply informed of God's love of and plan for the entire creation, they would amend their destructive practices. Local and diocesan committees sent hundreds of informational packets to parish priests, encouraging them to initiate "earth-friendly" study groups, sermons, and ministries. Over time it became clear that more information did not lead to transformation. To paraphrase Paul, human sin is rarely susceptible to cognitive cures.

As I studied further, it seemed that treating Christian-earth relationships in terms of discipleship or virtue might generate results that were both more productive and more faithful than either secular environmentalism or Christian education could produce. The result of this thinking became (in geologic time) a PhD dissertation and, eventually, this book.

I must acknowledge the religion faculty at Duke University for their model of outstanding scholarship and personal involvement in theological work, especially my advisor, Stanley Hauerwas. Stanley's responses to my efforts were usually encouraging, sometimes maddening, and often indecipherable, but he always, always made me think. This book would also never have come into being without the continued fellowship and guidance of Rev. Dr. N. Brooks Graebner and the people of St. Matthew's Episcopal Church, Hillsborough, North Carolina. They set a high standard for serious, good-spirited wrestling with issues of faith and ethics. Finally, my family deserves deep gratitude for supporting my years of work, even when they would have rather been doing something else.

CHAPTER 1

THE CHURCHES' RESPONSE TO THE "ENVIRONMENTAL CRISIS"

L ET US begin with an image, an eschatalogical vision. Imagine driving to an unfamiliar church on Sunday morning, a church you have heard is worth visiting. When you are a few blocks away, you begin to notice a number of people walking and bicycling in the same direction you are headed. They look not like athletes or fitness buffs, but like families on their way somewhere. Turning onto the church's street, you begin to hear the church bells, and see the bikers and pedestrians increase their pace. Finally you see the sign for the church, but you cannot actually see the building—just trees and bushes and a very small gravel parking lot. A church bus pulls into the lot behind you, and you grin at its "Biodiesel for Jesus" bumper sticker.

The passengers get off the bus, greet you warmly, and lead you along a path from the parking lot to the building. Along the way they explain that the large garden you are crossing grows organic fruit, vegetables, and flowers for local soup kitchens and church altars. Members of the church, as well as the local neighborhood, tend the garden and compost bins. They tried, they tell you laughingly, to produce their own communion wine, but all they managed to make was purple vinegar! So they buy their organic wine and communion wafers from elsewhere. The garden, however, serves as a project in which the whole community can participate, and it makes good use of land that had formerly been a parking lot. As your guide says, "We take literally the phrase 'taste of the kingdom!'"

Finally you emerge from the rows of corn and sunflowers. The church stands before you, constructed of a combination of local stone, recycled wood siding, and other "green" materials. It looks much like any other church, although the roof is nearly covered with solar panels. Inside, the sanctuary and meeting rooms glow with natural light, and you feel the movement of air from fans and open windows. The sanctuary is arranged in a contemporary style with the altar placed so that the minister or priest faces the people during the service. Behind the altar and cross, a huge glass window looks out onto what appears to be a park or meditation garden—trees, shrubs, birdhouses, and a fountain. It is a rich visual scene and it is very different from churches where no hint of the "natural world" is permitted to enter.

The service itself strongly resembles that of your home church, although it includes prayers for the local river (site of a recent toxic spill) and for endangered animal and plant species. The preacher focuses on hope, and the privilege and responsibility of Christians to bear witness to the eschaton—the fulfillment of God's promises for creation—as envisioned in Isaiah's and Jesus' analogies of the Kingdom. After the service, an older man offers to show you around and tell you about the church's various ministries. These run the gamut from missionary efforts to racial reconciliation to assisting a congregation in post-Katrina New Orleans, but several focus on ecological or eco-justice concerns. The church buildings are extremely energy efficient, using combinations of geothermal and solar energy. Each household is invited to make an eco-justice covenant, in which it commits to responsible consumption, patronizing fair-trade businesses, and reducing its ecological "footprint."[1] In some ways this church seems idealistic or even naïve about its ministry goals. When you try to ask your guide about this, he laughs. "I suppose we might seem naïve to a visitor, with our grand schemes to reduce carbon output, clean up the river and provide healthy food for homeless folks. But we don't think we can fix the world. We just like to show the possibilities of what

1. An ecological "footprint" is a way to estimate, based on things like miles traveled per year and size of house, "how much productive land and water you need to support what you use and what you discard," http://www.earthday.net/footprint/info.asp (accessed October 18, 2007).

Christ can do—and Christ never worried about practicality, did he? In fact," he continues, "what we've found is that improving the way we live on the land is a process. We take a step that seems big, then we find it wasn't so hard after all. So we take another step. Pretty soon, the church is engaged in what some might regard as "radical environmentalism." Yet, most of us don't think of ourselves as environmentalists, really. We're just Christians, trying to be faithful as best we can."

⁂

This is a fantasy church, a product of one person's hope, theology, and imagination. The following chapters will argue that holding such an eschatological vision of church—so long as it is guided by scripture and theological tradition—is a critical aspect of Christian discipleship. Christians cannot testify to something we cannot imagine, even if we know that our eschatological imagination can only offer a dim, cloudy impression of God's eschaton. However, many real churches in America (and elsewhere) share a similar vision, and they are enacting bits and pieces of their ideal, as the following examples demonstrate.[2]

First United Methodist Church in Porterville, California, owns a half-acre, irrigated organic garden. Plots are available to rent for a small fee to cover water usage. "The mission of First United Methodist Church Community Garden is to enhance the quality of urban life and strengthen cross-cultural community bonds by creating and sustaining an organic garden that will promote sharing healthy food, environmental stewardship and recreation."[3]

Similarly, the Fourth Presbyterian Church in Chicago has developed an organic community garden on property in the Cabrini-Green community, an area long noted for poverty, crime, and urban social problems.

2. These churches would not necessarily agree with the eschatological argument of this book. If their ethical framework is more of a divine command model, they believe it is important to care for God's creation simply because God desires or commands that we do so. Nonetheless, they serve as examples of what any church can achieve, with the grace of God and the power of the Holy Spirit.

3. First United Methodist Church, Porterville, California: http://www.gbgm-umc. org/porterville/garden.htm (accessed October 18, 2007).

As mixed-income housing is replacing the public housing buildings, the neighborhood is undergoing dramatic change and upheaval. Fourth Church is dedicated to supporting local residents and strengthening ties between its members and those of Cabrini. In cooperation with Growing Power, a leader in urban agriculture, and Greencorps, a Chicago city project, the church has established an organic garden for all community members to use. The garden includes both individual and community food plots, a butterfly garden, and native plant garden. Church members and neighborhood volunteers work community plots, and the food is given away to individuals who request it and to a Fourth Church meals ministry.[4] Community members participate by gardening, visiting, engaging in children's programs, cookouts, and other social events. Giving and receiving food from the garden is only a small part of the interaction between Fourth Church members and the larger community.

Fourth Church emphasizes to the gardeners that,

> while we take pride in the delicious vegetables and beautiful flowers we can grow on top of these former basketball and tennis courts, our real focus is in growing relationships between the members of Fourth Church, the residents of Cabrini-Green, and the residents of the neighboring condos and apartments. Your willingness to extend an open hand and open heart to each other is the ground from which community can blossom.

These efforts are bearing fruit; participation has gradually increased among both children and adults. Anne Harris, Program Manager, reports,

> people have shared with us on their visits how much it means to them that we are doing this—they appreciate the beauty of the place, that we are taking good care of the children, that we share our bounty—and we've heard a number of comments that we are blessed, in that we have not had any damage or serious vandalism done to the site, despite its location and the problems on the other side of the fence.[5]

4. Anne Ellis, e-mail message to author, June 5, 2007.
5. Anne Ellis, e-mail to author, June 6, 2007.

Through the garden and associated programming, Fourth Presbyterian Church and Growing Power model a rich integration of earth ministry, eco-justice ministry, and social ministries.

With regard to energy conservation, many churches have replaced their incandescent light bulbs with fluorescent ones, and they have strengthened their building's insulation. Madison Christian Community in Wisconsin, a partnership between Advent Lutheran (ELCA) and Community of Hope (UCC), took several steps farther. In 1992 MCC installed photovoltaic cells on the roof of the church building. "In installing the unit, the Madison Christian Community hopes to reduce energy consumption and emission into the atmosphere, and be a visible example to the general public of their commitment to stewardship of the environment."[6] The community tracks power production and usage by computer, so it can monitor high energy-use activities. Madison Christian Community is not focused solely on energy, however. Since 1983, the two churches have restored a small prairie on the grounds. "As part of our mission to care for our earth, this prairie preserves a part of the natural diversity that thrived before European settlers came to this part of southern Wisconsin. It provides cover for small mammals, insects, and birds that are losing their habitat as the west side of Madison develops."[7] Restoring and caring for the prairie requires significant effort. Native seeds were collected from along railroad tracks and steep hillsides, and the prairie is burned every spring to eliminate invasive species and mimic the natural fire-germination pattern of growth. Finally, the MCC has set up a lender program for household items that receive only occasional use: rototiller, air mattress, and power tools. This program not only discourages excessive accumulation of household possessions, but it also encourages fellowship by creating additional connections between congregation members.[8] This is a simple, cost-free program that can make a difference in the life of the church and in its ecological ministry.

6. Madison Christian Community, http://www.madisonchristiancommunity.org/environmentalstewardship.htm (accessed October 18, 2007).

7. Ibid., http://www.madisonchristiancommunity.org/prairie.htm (accessed October 18, 2007).

8. Ibid.

Saint Matthew's Episcopal Church in Hillsborough, North Carolina, is one of few that have addressed the issue of the Eucharist in the context of ecological destruction. After considerable time spent in research and feasibility study, the church now uses only organic wine and bread from small, labor-friendly sources for communion. Members can now partake of "the gifts of God for the people of God" with less reliance on harmful pesticides, herbicides, or oppressive labor conditions. Of course, the church also serves only triple-certified coffee (organic, shade-grown, and fairly traded).

One of the most impressive examples of "green discipleship" is the Georgetown Gospel Chapel in southern Seattle, Washington. This small, "full-gospel" church is located in a historic neighborhood near a Superfund toxic-waste site and Boeing field. As "a tiny enclave of homes and businesses hemmed in by factories, warehouses, freeways, railroads, barge terminals, and airplanes," Georgetown continually struggles to defend its livable character against industrialization and toxic dumping.[9] Twenty years ago, the church replaced its lawn (and expensive irrigation system) with a flower and produce garden. The garden is watered through a water reclamation system that reduces runoff into the Duwamish River. Pastor Leroy Hedman and his parishioners offer gardening and composting expertise, seeds, and tree seedlings to members of the local community, as well as food from the garden. The after-school program for local children includes instruction and experience in caring for creation, and adult education includes information about local toxic-waste sites and the status of their reclamation. The Chapel has also instituted several energy-saving measures with more efficient lighting and appliances and better insulation. In 1999 Georgetown Gospel Chapel was the first congregation to win an EPA Energy Star Award.[10] The money saved through energy conservation is directed toward outreach and mission activities.

9. Washington State University: http://www.historylink.org/essays/output.cfm?file_id=2975 (accessed October 18, 2007), and save Georgetown: http://www.savegeorgetown.org/Home/tabid/36/Default.aspx (accessed October 18, 2007).

10. Earth Ministry, http://www.earthministry.org/Congregations/stories/Georgetown/Georgetown.htm. See also Environment Protection Agency, http://www.energystar.gov/index.cfm?c=sb_success.sb_successstories_georgetown (accessed October 18, 2007).

"It honors Christ to serve the creation," Hedman told *Charisma*. "Why should the New Agers be the ones gathering the attention for preserving 'Mother Earth'?"[11]

Other churches, such as Saint Mary's Catholic and First United Methodist churches of Corvallis, Oregon, hold "Meet Your Farmer" socials. At these events church members initiate personal connections with local farmers, learn about their struggles and successes, and purchase shares of local crops via "community supported agriculture" (CSA). CSA not only supports small farmers, but it also helps establish relationships between growers and consumers, shortens food travel, provides fresh produce to consumers, and helps preserve farmland.

These examples are not churches with unlimited resources (Georgetown Gospel Chapel has only a few dozen members), or churches that have substituted ecological diatribes for the Christian gospel. Rather, they view ecological and eco-justice issues as central to Christian life. The question is, why are these examples the exception rather than the rule? Perhaps some recent history will help answer that question.

In 1968, two years before the first "Earth Day," the Lambeth Conference of the worldwide Anglican communion both welcomed "the prospect of human control of the natural environment," and, somewhat conversely, urged all Christians "in obedience to the doctrine of creation, to take all possible action to ensure man's responsible stewardship over nature; in particular in his relationship with animals, and with regard to the conservation of the soil, and the prevention of the pollution of air, soil, and ocean."[12] Since then, every Lambeth Conference and every General Convention of the national Episcopal Church has called for increased attention to, and correction of, ecological destruction and environmental racism.[13] These calls join a chorus of official statements and theological

---

11. Adrienne S. Gaines, "Christians Urged to Care for the Earth," *Charisma* (September 2004), http://www.charismamag.com/display.php?id=9580 (accessed October 18, 2007).

12 Coleman, *Resolutions of Lambeth*, 152–53.

13. Episcopal Church of the U.S.A., Resolution 2000-D005, http://www.episcopal-archives.org/cgi-bin/acts_new/acts_resolution-complete.pl?resolution=2000-D005 (accessed March 15, 2005).

treatises from nearly every Christian denomination in the U.S.[14] Yet, notwithstanding the examples above, most churches' ecological discipleship remains overwhelmingly words on paper rather than significant works in practice. Many churches recycle their office paper and drink containers; some have switched from disposable cups to mugs for coffee hour. But very few, for example, have instituted any serious energy- or water-saving measures; few have begun carpooling to services, abandoned the use of toxic chemicals in maintenance of buildings, grounds, or linens (all those white vestments and altar cloths use lots of chlorine bleach!), or committed to buying bread and wine (or grape juice) from non-exploitative sources. While such small steps as recycling are important and laudable, they pale against the churches' own calls to resist "unjust distribution of the world's wealth, social injustice within nations, the rise of militarism, and irreversible damage to the environment."[15]

The churches' feeble response to this deep-seated problem is a likely consequence when creation and Christian ethics are regarded apart from eschatology. Christians need to re-view the biophysical universe as part of God's continuing creative activity that will culminate in the Kingdom, and Christian life as a communal testimony to that Kingdom. Such a re-conception is as much praxis as theory, as much traditional virtue as new mission: it will enable churches to undertake difficult, practical reforms of their life on earth for the sake of ecological discipleship.[16]

14. Official statements can be found at the web sites for each denomination's national office: Episcopal Church at ecusa.org, Lutheran Church at http://www.elca.org, United Methodist Church at http://www.umc.org, Presbyterian Church at http://www.pcusa. org, and Southern Baptist Convention at http://www.sbc.org. For useful analyses of these statements, see Fowler, *Greening*, and Van Houtan and Pimm, "Christian Ethics."

15. Resolution 40, 1988 in Coleman, *Resolutions*, 219.

16. The "we" in this book refers to Christians who share concerns about the damage to God's creation, but who find it difficult to adopt and implement changes in favor of 'earth friendly' practices. The assumed readers, then, are not Christians who are not yet convinced of the centrality of eco-discipleship in Christian life, although they may find some of the arguments interesting and helpful. Similarly, while this is an explicitly Christian theological work, some of its insights will benefit people with other—or without any—faith convictions.

This is a book in constructive theology. Therefore, it uses a wide variety of sources in sometimes surprising ways. It retrieves and recovers strands of the classic Christian traditions and weaves them together with the work of a wide variety of contemporary theologians and environmentalists. Theologians who might have deep disagreements on other matters are brought together for their combined ability to contribute to the argument that the church's most faithful mode of response to ecological issues is through Christian witness to the Kingdom of God. Rather than stopping to address in depth any one strand or contributor, I use these sources to articulate a reading of the tradition that demands the church's committed, energetic response to God's gifts and promises for creation.

This chapter examines the cluster of factors—political, cultural, historical—that has prevented the churches from more forceful engagement with ecological issues. It then addresses the theological factors that form the target of this book. Christian social ethics are commonly viewed as either ways Christians can contribute toward the management of social problems, or, in a less humble vein, ways Christians can build the Kingdom here on earth. Neither approach can sustain the challenges of faithful caring for God's creation. I then review the work of three prominent theologians concerned about ecological issues: Larry Rasmussen, Catherine Keller, and Rosemary Radford Ruether. Finally, I position this book in relation to those and other Christian eco-theological approaches.

WHY THE LIMP RESPONSE?

❧ What might prevent stronger environmental response[s] on the part of congregations? One factor is that environmentalists have focused their efforts on education as the way to promote eco-discipleship, as though the problem were one of convincing churches of the need to change. This is certainly true to an extent; many congregations, especially fundamentalist groups, have not seen ecology as a Christian theme.[17] But even "enlightened" congregations find the shift to eco-discipleship daunting. Human sinfulness, unfortunately, proves highly resistant to cognitive

17. K. S. Van Houtan and S. L. Pimm, "Christian Ethics," 124ff.

9

cures. We know this from Aristotle (*Nicomachean Ethics*, Book VII), from Paul (Rom 7:14–20), and most spectacularly from Augustine (*Confessions*, Book VIII.v). This is clear, too, from attempts in the 1960s and 1970s to "educate" white people out of racism. So it should not be surprising that massive efforts to educate the churches about ecology have borne little fruit in practice.

There have been no formal studies to date of churches' failure to "get green." But I suspect that the insufficiency of education is related to a combination of factors, some of which affect diverse ministries, and some of which are unique to ecological ministries. To begin with, the changes called for by church authorities—beyond recycling— are not easy, despite the attempt by environmental literature to portray them as "simple."[18] These changes are especially difficult for middle-class people who are strongly attached to the convenient and laborsaving routines their money and status have obtained. Consider, for instance, the difference between curbside recycling and gray water use for the churchyard.[19] Recycling requires simply placing bins near trash containers, publicizing their use, and designating someone—usually whoever already collects trash—to move them to the street on the appropriate day. Establishing gray water use, however, requires significant design and installation of a plumbing system (including the selection of people to run the project), changes in care of plantings, kitchen use, and laundry procedures. One or two people can manage the curbside recycling, but switching to a gray water system requires a much larger impetus by a group of church members (or a household) and a much larger outlay of money.

Secondly, authorities in mainstream white churches are relatively weak-voiced in contemporary U.S. culture. The very nature of the liberal church means that few congregations perceive the urging by bishop, supervisor, or national body to be sufficient reason to do anything— especially anything difficult, time-consuming, or unpleasant. As Stephen

18. See, for instance, The Greenhouse Crisis Foundation, *101 Ways to Help Save the Earth*.

19. "Gray water" is leftover water from dishwashing, drinking, and laundry that can be used to water plants. Gray water systems effectively allow almost two-for-one savings in water consumption.

Carter has argued, religious commitments in U.S. society are regarded with suspicion as soon as they conflict with secular mores. This suspicion is a natural outgrowth of the Enlightenment's construction of the rational, autonomous moral agent. Such an agent bears authority for his or her own life, so submitting to the authority of church leaders can only be done by choice or because of immoral coercion. In this context, choosing to let religious leaders determine significant aspects of one's life appears irrational or irresponsible.[20] So church teachings (in moderate and liberal churches, at least) become diluted from authoritative teachings into advice; following that advice is little more than a matter of preference.[21]

Moreover, the liberal church parish is seen as an aggregate of individuals, rather than as a body—specifically, the body of Christ. Thus the idea of the church reforming in accordance with the Gospel becomes diffused into the idea of multiple households and individuals reforming themselves. So we lose the importance of the church—both local and global—as a singular witness to the plenitude of the Kingdom through its ecological practices. Ecological transformation appears as a problem of "collective action," because the individuals who "attend" the church retain greater subjectivity, agency, or moral substance than does the church itself. It is interesting, in this regard, that many Americans find it much easier to speak of "nations" and "corporations" as subjects than of churches.

Thus far I have surveyed four causes of Christian churches' failure to respond to environmental crises: the ineffectiveness of education, the difficulty of implementing real change at the practical level, the relative powerlessness of church officials, and the common vision of a church as a collection of individuals, rather than as an active moral subject. Finally,

20. Carter, *Culture of Disbelief,* 23–43.

21. There are, of course, exceptions to this scenario. African-American churches and fundamentalist churches often invest clergy and church leadership with more authority over congregants' lives. And mainstream church leaders may flex their muscles for certain issues; Catholic bishops have excommunicated parishioners for torture in Latin America, and threatened to do so for abortion advocacy in the U.S. (Carter, *Culture,* 61; Cavanaugh, *Torture and Eucharist,* 105.) I do not believe an increase in ecclesiastical power would necessarily be a good thing, or relieve environmental harm by Christians. The cultural powerlessness of much church leadership, however, must be taken into account.

and most relevant to this project, congregations resist difficult environmental changes because the problem is posed as both overwhelming in nature and/or solvable by technology. On the one hand, congregational changes appear trivial against the global-sized task. In Wendell Berry's words, "How, after all, can anybody—any particular body—do anything to heal a planet?"[22] Christian environmentalists often fall into a despairing paralysis because no action by individuals or congregations or small communities can possibly make a difference. Why bother with the sweat and inconvenience of walking to church if oil consumption keeps rising so steeply? Why trouble to recycle if landfills keep growing? Why throw ice cubes in the ocean when global warming continues to increase?

At the same time, the "flush and forget" aspect of affluent life enables households and congregations to avoid confronting the results of their ecological practices. We tend to assume that technological expertise can—and does—cure any environmental disorder we might create. So perhaps trash in landfills does decompose into healthy soil, and waste poured down drains somehow transforms into farm fertilizer. Perhaps a sizeable segment of our energy bill contributes to renewable energy sources. Perhaps the pesticides used on grapes grown for our Eucharist wine are safer now for the farm workers. Perhaps the scientists and politicians, behind the scenes, render our church life relatively harmless to God's creation.

This two-sided construction of ecological questions as either unsolvable or already solved is, of course, contradictory and unsubstantiated. Yet, it holds great power as motivation and justification *not* to respond to ecological appeals from church leaders. If the environmental crisis is truly a global crisis, then a single church—or single collection of individual Christians—can do nothing to help. On the other hand, if scientists, economists, and government experts are producing solutions (as they often claim), then we can ignore the teachings of church authorities and environmentalists. It is crucial to note that both sides of this contradiction are strongly consequentialist in perspective: one's response to ecological damage depends on the anticipated consequences of one's

22. Berry, *What are People For?*, 197.

action. Action—particularly laborious, costly action—makes sense only if it will produce significant, desirable, and visible results. And if the church is perceived as relatively powerless in contemporary society, then its actions will never produce substantial results, except greater futility. This enthrallment with effectiveness pervades our society, but it has particular ramifications for Christian life. Some attention to effectiveness is, of course, part of Christian prudence; we want our deeds of charity and justice to realize our intentions. Too much focus on consequences, however, feeds into the notion that humans are God's managers rather than God's creatures—that only human efforts hold any hope for the future. In short, the Kingdom of God has moved from God's hands to human hands—from God's promise to the world, which we humans await and prepare for, to God's wish, which humans try to fulfill.

FROM THE KINGDOM OF GOD TO THE HUMAN PROJECT

❧ This is a relatively recent, yet enormous, shift in theology, with dramatic effects in every area of Christian life. It has happened so thoroughly and so broadly, however, as to be transparent. The "Kingdom of God" as something that humans try to create on earth is such a naturalized image that most Christians do not even realize it is a recent innovation. From the beginnings of the Church up through the Protestant Reformation, Christians disagreed vigorously about the details of judgment and salvation and the particular nature of God's Kingdom—who would be saved, the merits of faith versus works in the face of judgment, whether human relationships and activities would continue, what if any non-human creatures would be evident, and so forth. Nearly everyone, however, agreed that the Kingdom was something that God would bring about in God's own time; the simple, yet awesome task of humans was to prepare themselves so they might fully participate in the Kingdom when it arrived, and to witness to Christ as the inaugurator and Lord of the eschaton. Notable exceptions to this conviction arose from time to time, usually during a period of intense conflict or social change. So Thomas Münzer, in 1525 in Germany, led a violent uprising against the Lutheran-supported authorities in order to ignite the Apocalypse and to establish

God's reign on earth. And Gerrard Winstanley's Diggers in England in the seventeenth century defended the principle of common land through subversive farming, declaring that God intends for land to be shared among all. "True religion and undefiled is to let everyone quietly have earth to manure."[23] Once land was distributed justly, Winstanley argued, the Kingdom would naturally prosper.[24] Münzer's utopian revolt was violent (one hundred thousand peasants were killed in the 1525 Peasant Revolt), while Winstanley's was peaceful; nonetheless, neither succeeded. The point, however, is that these exceptions are so scarce in church history as to prove the rule.[25] How is it, then, that many (if not most) liberal Christians now assume that the Kingdom is a human project, something we make rather than proclaim?

In a highly simplified summary, here is what seems to have happened to eschatology in modernity (and Immanuel Kant, of course, stands at the center).[26] In the wake of seventeenth-century violence, deists and rationalists emphasized the reasonableness of religion. The essentials of faith were those that everyone could agree to; any sort of privileged information—such as a biblical description of the Kingdom of Heaven—was unnecessary speculation and probably best discarded. Kant, although opposed to Rationalism on many fronts, took a similar stance with regard to religion. In Kant's short essay "The End of All Things" (1794), he argues that the idea of a Final Judgment persists because the moral progress of the human race lags behind its progress in talents, art, and taste, so people suffer the perpetual sense of being morally inadequate—even though the human race is more advanced morally than ever before, and continues its

---

23. Quoted in Keller, *Apocalypse Now and Then*, 194.

24. Taylor, "Gerrard Winstanley's Theology," 105–6. Winstanley is a fairly obscure radical from the English Reformation; nonetheless he has a statue in Moscow's Red Square, and has been recently denigrated in the West as a despot imposing his utopian vision on others.

25. These examples also point out how belief in God's Kingdom can work for or against established authorities, depending on the prevailing political and religious climate. In our own times, human attempts to establish (or "co-create") the Kingdom almost inevitably lead to collusion with state and corporate imperialism.

26. I am indebted to Reinhard Hütter for suggesting the outline of this history.

development in this arena. Similarly, the idea of the Kingdom provides a completion, a resting-point, to humanity's vision of the slow process of moral development, "a contentment which he can think only by thinking that the ultimate purpose will some time finally be reached."[27] Therefore, since the ideas of Judgment and the Kingdom result, in essence, from a perceptual difficulty, and since whatever happens afterward is not theoretically conceivable to us, these ideas "should be regarded rather in a practical sense, not laboriously pondered with respect to their objects . . . but rather as we are required to contemplate them on behalf of the moral principles which pertain to the ultimate purpose of all things."[28] Kant goes on to say that the image of the Final Judgment should not motivate us to do good deeds out of fear or ambition, but that it is only "a generous warning arising out of the legislator's [God's] good will, to be on our guard against the damage which should inevitably spring from the transgression of the law."[29] Note what underlies this argument: Kant retains the idea of a telos ("ultimate purpose") to the world, which he sees as human moral perfection. This telos arises from the nature of the universal moral order that exists, in a sense, even prior to the Creator. For the Creator has no choice regarding the consequences of obedience to, or violation of, the law. But humans are continuously progressing in their moral development, and such progress seems independent of God. Christianity, and Jesus Christ himself, bears only a pedagogical function once the world is created. Even this, however, seems somewhat superfluous because human moral progress is so inevitable, in Kant's view, that the lack of one teaching tool or another would not matter much. Thus we have a strong progressivist view of history combined with a reduced view of God's activity in the world. Kant does not quite declare humans to be the builders of perfection; rather, his emphasis in this essay is that we should console ourselves that our imperfections will eventually diminish.

Now we jump forward about a century. The narrowing of theological topics to ethical concerns continued among liberal theologians, especially

27. Kant, *End of All Things*, 79.
28. Ibid., 76
29. Ibid., 82.

in Germany. Albrecht Ritschl, a liberal German theologian (1822–1889), described Jesus' Kingdom as complete communion with God, the telos toward which human moral development progresses. But this development is not automatic or inevitable even though the Kingdom exerts a continuous forward tug on history. It needs encouragement and support by godly people—in particular, the Christian community, whose obedience to God's will makes the community a fitting helpmate for the Kingdom. Walter Rauschenbusch, a Protestant theologian (1861–1918), studied with Ritschl before ministering in New York's "Hell's Kitchen" area. In *Christianity and the Social Crisis*, and in other works, he declared that it is the *responsibility* of humans to institute the Kingdom of God on earth by organizing society according to Gospel precepts, to "Christianize the social order." "The future of Christianity itself lies in getting the spirit of Jesus Christ incarnated in history."[30] So the shift from faith in human progress to administration of human progress seems complete.

This is an interesting and deceptively powerful theological turn. Each of these three figures worked from benevolent intentions, and Ritschl and Rauschenbusch carried out insightful analyses of social and economic injustice. Yet, part of their legacy is this notion that humans—in fact, a very select group of humans—are duty-bound to enact heaven on earth, to transform society into (their view of) the heavenly Kingdom. History co-opts eschatology. For both Ritschl and Rauschenbusch, the church functioned as the "change agent," the unique medium through which the Kingdom would be realized. In later decades, however, as Social Gospel ideas of a good society gained currency in America, the distinctiveness of the church dissolved. In Stanley Hauerwas's words, "it is by no means clear why you need to go to church when such churches only reinforce what you already know from participation in a democratic society."[31] So the Christian task of establishing the Kingdom began to appear essentially the same as the secular task of constructing a just society. Christianity, moreover, becomes an instrument or consultant for this task, rather than a story about God's judgment of all such kingdom-

---

30. Rauschenbusch, *Christianizing the Social Order*, 127.

31. Hauerwas, *Better Hope*, 26.

building projects. Thus, in this view normative statements are judged first on their consonance with ethics and only secondly on their consonance with Christianity.

If the Kingdom, in its secular or Christian versions,[32] is a human project, then efficacy and expertise become more important than faithful obedience. That is to say, if humans (more specifically, our particular affluent, enlightened subgroup of humans) are the overseers, then we must rely on technical expertise to move forward. On the other hand, if technical expertise fails, *our* powerlessness entails insolvability: if humans cannot do it, it cannot be done. So we slip into the trap between managing an environmental crisis through existing science, market economics, and traditional politics—in which the church occupies only a minor advisory role—and succumbing to an overwhelming environmental crisis through helplessness and despair—in which the church plays no role at all.

Put this way, the contrast with biblical narratives jumps out. After all, two of the central themes of the biblical story are: first, God's ability to do the impossible (the flood, dividing the Red Sea, giving a child to a barren woman, resurrecting the dead); and second, the futility of human ambition to be like God or to do what God does. This message emerges at the very beginning of the Bible. In Genesis 1–3, it is God who creates the heavens and earth, and the humans who suffer the effects of trying "to be like God, knowing good and evil" (Gen 3:6). And this theme continues throughout the stories of God's relationship with God's people. God can command a great flood, divide the sea, give children to barren women, and even resurrect the dead. People, on the other hand, come to grief whenever they imagine themselves outside the limitations of creatureliness: consider the tower of Babel or the arrogance of King David (e.g., 2 Sam 11, and 1 Chr 21:1–6). "The human race has nothing to boast about . . . If anyone wants to boast, let him boast about the Lord."[33]

Believing either that environmental damage can be undone by existing political-economic systems, or that it cannot be halted at all, and

32. I have placed secular and Christian together here because, for many Americans, "heavens" look very similar, with variations in the presence of Jesus or sex, perhaps. Many pious Victorians anticipated "free love" as a key part of heavenly delights.

33. John Yoder's paraphrase of 1 Cor 1:29–31. Yoder, *He Came Preaching Peace*, 46.

acting on those beliefs, is to imagine that we are beyond any dependence on God, and so to depart from the Christian models of faith and obedience. Neither faith nor obedience can stand alone; each requires the other. Faith alone that God will rescue us and rescue the earth leads to an abdication of responsibility for our own behavior, or even an apocalyptic sensibility that the current horrors of ecological disasters signify the imminence of God's coming.[34] Obedience alone, however, like that of the grim stereotype of colonial Puritans or the Lutherans in *Babette's Feast*, abandons the joyous praise of God's work and hope in God's promises for the future.[35] Faithful obedience, however, means life carried on in accordance with God's commandments in confidence that God's will triumphs even if ours does not. Thus we live in witness to the good news that even though humans are not in charge—of history or nature—the Kingdom has already been inaugurated and will be consummated in God's own time.

AN ALTERNATIVE TO EFFICACY

So the modern focus on efficacy is doubly flawed: it distracts Christians from central eschatological truths, and it hinders what should be Christians' hopeful, active response to calls for ecological reform. The trouble is both theological and practical; or more properly, it demonstrates how theology and practice are not really separable. What would happen, then, if attention shifted from effectiveness to eschatology, from what we can accomplish to what God has already accomplished in Jesus' resurrection? The following chapters argue that careful attention to rig-

---

34. See, for instance, Lindsey, *Road to Holocaust*, 201, 218. The identification of ecological destruction with signs of the end times is common among certain strains of Christian fundamentalism. The response called for is not environmental activism, but "a personal invitation of Christ into the heart and an acceptance of the gift of forgiveness which He gave His life to provide," (Lindsey, *Late Great Planet Earth*, 149). Accepting Christ will not avert global environmental catastrophe, but will save oneself from its deadly effects, as believers will be taken up with Christ before the earth is destroyed. It should be noted that not all fundamentalists follow this end-times scenario. Fundamentalism does not always correlate with disregard for environmental protection.

35. Betzer, *Babette's Feast*.

orous eschatology enables the following for Christian ecology: (1) the recovery of a strong creation-redemption link through Jesus Christ; (2) a more proper understanding of human efforts as witness to, rather than establishment of, the Kingdom of Heaven; (3) a critique of the race, class, and gender assumptions underlying dominant eschatological visions; (4) and the development of hopeful patience as the means to continue eco-discipleship in the face of apparent failure.

It may appear that I am promoting a sort of quietism, an eschatological expectation in which we fold our hands in our laps and await God's heroic rescue of the world, as we sigh over the continued presence of evil. Instead, I want to question, in line with John Howard Yoder, the presumed choice between ethical management through existing socioeconomic systems and passive accommodation. Jesus Christ refused the possibility of enacting change through worldly power: he rejected armed rebellion against the imperial Romans as well as accommodation with them through dominant Jewish factions. In the crucifixion, he both demonstrated and inaugurated the "way of the cross" by which nonviolent witness to the kingdom *is* God's acting in the world, which has already triumphed in the Resurrection. "Our readiness to renounce our legitimate ends whenever they cannot be attained by legitimate means itself constitutes our participation in the triumphant suffering of the Lamb."[36] Thus obedience—faithful living as the body of Christ (the church) in witness to the Kingdom—is more important than efficacy. We do not have to "win," because God has already won. Therefore the activity (or activism) of the church should be a continuous, visible demonstration of radical alternatives made possible by God in Jesus Christ. (The crucial question of what this looks like in regard to ecology will be explored in chapter 4.)

"Witness work" may not look different than "effective work" in all cases; to a certain extent, recycling is recycling is recycling. Whenever ecological efforts, however, are constrained by the language of "practicality" or the question, "what difference would it make?" the discourse has shifted to what we can do in the secular world, rather than to what God

---

36. Yoder, *Politics of Jesus*, 244.

makes possible in the world ruled by Christ. Moreover, continued focus on God's work in the Kingdom facilitates the nourishment of our lives by the resources of our own tradition, including teachings of church authorities, historical examples of ecological living, liturgical formation especially through Baptism and Eucharist, and theological understandings of Jesus Christ as creator and redeemer of all. Joseph Sittler says of Christ (paraphrasing Col 1:15–17): "He comes to all things, not as a stranger, for he is the first-born of all creation, and in him all things were created. He is not only the matrix and prius of all things: he is the intention, the fullness, and the integrity of all things: for all things were created through him and for him."[37] Proper worship of this Christ entails the difficult, communal labor of orienting our congregations toward ecological discipleship, in witness to the infinitely abundant Kingdom of Heaven.

This is a "boldly apologetic" work. By this I mean, first, that it does not attempt to re-make Christian doctrines and traditions, but to interpret them in the light of current conditions. Secondly, however, it moves beyond the usual categories of "stewardship" and "dominion," and the commonplace dicta that individual Christian virtue is sufficient for Christian ecology. What is often forgotten today is that during most of Christian history, the fundamental ontological distinction lay not between humans and nonhumans, but between God and everything else—God's creation. Thus I choose to focus not so much on the question of humans' role vis-à-vis other kinds' (though that question never completely disappears from the discussion), but on God's plan for all of creation.

ROUTES NOT TAKEN

❧ In contrast to this apologetic orientation toward Christian traditions, several of the most popular and prolific writers in ecospirituality work from the assumption that Christianity is ecologically culpable.[38] That is

37. Sittler, "Called to Unity," 177–78.

38. Thomas Berry, Larry Rasmussen, Catherine Keller, and Rosemary Radford Ruether are the primary examples of this trend. All of them follow (at least partially) Lynn White's famous argument. White, "Historical Roots of Ecological Crisis," 1203–7.

to say, the history of Christianity—specifically, the history of Christian people since the Enlightenment—is so deeply implicated in patterns of environmental destruction that the faith itself is to blame. This interpretation holds that this destructive history was bound to result from adherence to central Christian doctrines. Consequently, Christianity itself, or significant parts thereof, must be abandoned in favor of an "earth faith" or ecological cosmology if the earth is to be saved.

CHRISTIANITY AT FAULT?

Rosemary Radford Ruether, for instance, began analyzing the connections between Christian doctrine, Christian practices, sexism, and ecological damage well before most other theologians. What Ruether finds most damaging in Christianity is a legacy of alienation from, and domination of, the rest of nature. The concept of an utterly transcendent, disembodied God who relates by command has promoted an other-worldly spirituality in which closeness to God requires escape from biophysical materiality. Further, "man" being created in God's image solidifies the distinction between humans (primarily male humans) and other nature into an absolute dualism, whereby evil and sin are identified with the natural. Ruether places the blame for Christianity's faults primarily in two places: early Christianity's adoption of ancient Greek (Platonic) ideas, and the formulation of doctrine by elite men. [39] Yet, Ruether's picture of historic Christianity is incomplete, because it leaves out the incarnation—*the* pivotal event of Christianity—through which Christians know God to be *not* disembodied, utterly transcendent, or removed from the world.

In a different move, Catherine Keller looks within Christian history for exceptions to Western Christianity's apocalypticism. She argues in *Apocalyptic Now and Then* that apocalyptic dualism runs as a destructive underground current through Christian history, leading to either violent domination of the "other" by whoever see themselves as God's army, or passive pathetic waiting for divine rescue from the consequences of our own behavior. Part of the apocalyptic legacy is the opposition between human and nonhuman, sacred/spirit and divine/flesh. So the

39. Ruether, *Gaia and God*.

bulk of Christian traditions, for Keller, has produced a history of ecological, social, and economic imperialism and oppression. However, Keller's own research reveals several historical examples of "counter-apocalyptic" Christianity, including the late thirteenth-century Beguines, seventeenth-century Diggers in England, the Shakers in America, Pierre Proudhon, and Saint-Simon.[40]

Other writers focus their criticism on the ecological destruction wrought by Christian colonialism in the Americas, Africa, and the Caribbean. Yet, the recent Christian history they point to, rightly, as being a record of appalling consumption, destruction, and waste of the natural world is, in fact, a history of only white Euro-American Christians, and even a partial history at that. Christian histories outside this dominant stream, such as monastic group life, counter-cultural religious communities, and American slave Christianity tend to be ignored.[41] Second, as Loren Wilkinson notes, the construal of Christianity presented as the target is itself a product of the Enlightenment more than of the longer history of theology.[42] The God who is distant and uninvolved with "his" creation is the God of Kant and Thomas Paine, but not much like the God of Abraham and Paul.

Certainly it is crucial for Christians to face the devastation their own history has left upon the planet, even if that devastation is not the whole history. Two responses are then possible: to emphasize (as I do) the failure of Christians to fulfill their own faith commitments regarding earthly life, or to seek an alternative to Christianity—or a revision of Christianity—that is more "earth-friendly." The latter response entails a blurring of the particularity of Christian doctrine and discipleship. In this case, Christianity is viewed as one among a number of religious traditions, each of which is a basket of tools that may be picked up or discarded at will, according to their usefulness for the larger task of sustaining the earth. Such a view, though, construes Christianity and other faiths in

40. Ibid., 113, 206.

41. Catherine Keller's *Apocalypse* cites excellent examples of "marginal" Christians doing sustainable ecological work, such as the Beguines, Quakers, and Diggers. Other examples can be found in Latin American spirituality and African slave spirituals.

42. Wilkinson, "New Story of Creation," 26–36.

modernist fashion, as "voluntary associations" relegated to the sphere of private, individual belief. If we can select our religion the way we select our exercise regimen, it can hold no power over our lives contrary to our own momentary decisions. In Ronald Beiner's words,

> The liberal vision of the individual as the autonomous chooser of his or her own purposes presupposes that the chooser is sufficiently sovereign over, and therefore distanced from, the choices that compose his or her identity that none of them must be regarded as binding. However, this conception of the self is incoherent, for a self that is as open-ended as the liberal conception requires would be not so much free as identity-less. Only a "thickly constituted self" shaped in its very being by traditions, attachments, and more or less irrevocable moral commitments can actually make choices that count.[43]

Despite this incoherence, liberal eco-theologians often regard Christians not as members of the body of Christ, but as individuals standing outside the traditions, or perhaps choosing to stand within a tradition for a time. Christianity becomes an optional affiliation and the church becomes a purely human object rather than a subjective agent of God.

"Earth Ethic"

One example of the "earth faith" trend is Larry Rasmussen's book *Earth Community, Earth Ethics*. Rasmussen's thesis is "our most basic impulses and activities must now be measured by one stringent criterion—their contribution to an earth ethic and their advocacy of sustainable earth community."[44] He wants to construct, recover, or synthesize a new or different religious faith that can serve the interests of a beleaguered earth. He is willing to do this from whatever resources are available and effective—where "available" means credible for modern, scientifically-friendly Western people, and "effective" means motivational in the direction of

43. Beiner, *What's the Matter*, 16.
44. Ibid., xii.

ecologically sound spirituality and practice.[45] Such an earth-centered faith will accord with Christianity in some respects, and clash in others. In fact, Christianity for Rasmussen is less an ongoing tradition that creates and surrounds us than a "thing" outside of us, a possible store of earth-faith ingredients.[46] To describe Christianity as a collection of tools or ingredients for a larger, more relevant project, however, is to render the beliefs and practices of most Christians unintelligible. Christian faith and ethics are not so distinguishable that Christian ethics might be described as environmentally useful while the basics of faith are disregarded as irrelevant or even false.

Rasmussen's description of Jesus is an example of the tendency to generalize Christian teachings. At this point in his book, Rasmussen has elicited/constructed an earth ethic based on creation's revelation of the divine. Now he argues in addition that such an ethic is inadequate if it omits attention to suffering in the natural world—the prevalence of great pain, distortion, and blight among all creatures.

> The kind of earth ethic that reads the presence of the divine from the flourishing of nature on its better days, or that reads the presence of the divine off the beauties of nature as we observe or imagine those (e.g., Job 12) speaks of God in true but limited ways. God *is* present in creation's beauty. The holy most certainly *is* mediated in nature's grandeur. But God is also present in the crosses of pain and twistedness and whatever other ways by which creation is violated.[47]

Instead, the revelation of God through Jesus Christ models the compassionate presence of God in the midst of suffering and despair. "Jesus is the

45. Some of his recent works are: "Redemption," 10–11; "Global Eco-Justice," 515–30; "Integrity of Creation," 161–75; and "Toward an Earth Charter," 964–67.

46. Rasmussen, however, does not quite deny the subject-creating power of Christianity. Indeed, it is because Christianity has created us as earth-aliens that it must be relativized under the primary goal of earth-faith. Rasmussen wants us to re-create (re-interpellate?) ourselves as stewardly earth-creatures. It is not clear why his invitation would appeal to any self-conscious person who is already opposed or indifferent to environmental goals.

47. Rasmussen, *Earth*, 286.

single most definitive revelation" of God, though not the exclusive one.[48] This Jesus directs our attention and energy toward compassion, "suffering with," as the starting point for earth ethics.

> The quest of a religious earth faith in this understanding is precisely for a power that overcomes suffering by entering into it and leading through it to abundant life, with abundant life pictured as the Sabbath condition of redeemed creation. God's goal is newness of life; but God's means is overcoming by undergoing; and God's way—to recall Luther—is seen in *Jesus as a living parable* of God's compassion in human form.[49]

For Rasmussen, the point of Jesus seems to be pedagogical and epistemological, rather than ontological. That is, despite the reference to power overcoming suffering, the picture of Jesus in this quote functions as a moral exemplar rather than Lord of all creation.[50] The voice of resurrection is quite muted here. Instead of creation's origin and redemption occurring by and through Christ, the possibility of redemption is diffused throughout creation: "The issue for earth ethics is the discovery of a power throughout creation that serves justice throughout creation . . . It is a matter of returning to our senses both to know the good and to do it, with compassion for earth's suffering as the single strongest sense."[51] This approach to redemption implies two related things: first, God's saving work is not merely unfinished, but somehow failed; and second, Christ is not so much the agent of salvation as a visible locus of this broadly dispersed redeeming power. However, Christians worship not "a living parable of God's compassion," but "one Lord, Jesus Christ, the only Son

48. Ibid., 285.

49. Ibid., my italics.

50. "Moral example" theories of the atonement are not unknown to Christian tradition, with Peter Abelard as the most famous proponent. Abelard, however, wrote about the atonement in the context of a rich understanding that Jesus was the Son of God. Rasmussen, on the other hand, only discusses Jesus in relation to the revelation of the cross. Because he does not mention any other aspect of christology (or Trinity), it seems that the whole point of Jesus is the teaching of the cross.

51. Rasmussen, *Earth*, 292.

of God." It is hard to see why such worship would not be, on Rasmussen's account, not only superfluous, but also misguided.

Also implied by Rasmussen's goal of earth faith is that "earth" is understood as prior to, and apart from, the faiths of its people; earth has its own story that might not match the story told by any particular religion. Nature is construed as self-interpreting, and Rasmussen is quite sympathetic to recent biophysics that emphasizes the continuity of all life. "Science's contribution is a story of origins: the stunning portrayal of a common creation in which we are radically united with all things living and nonliving, here and into endless reaches of space, and at the same time radically diverse and individuated . . . ."[52] Of course this "story" is itself a reading, an interpretation, as much as was Bacon's vision of nature as a female body to be bound and probed for her secrets.[53] This is not to say that scientific—or any other—interpretations are all equally valid or beyond appraisal. However, to presume that a historical moment in science stands in judgment over theologies, as though the former were truth and the latter merely legend, is a mistake. It may be that contemporary science tells Christians more about our own story, brings more of God's grace to the forefront, as it were, but that is because we *already* frame those scientific findings within the relationship of Creator to her creation.[54]

In Rasmussen's account, then, earth has a story that has taken a disastrous turn, largely due to Western human misbehavior, which is produced by faulty narratives in misguided traditions. We have, through our stories and beliefs, construed ourselves as utterly separate from a mechanical universe. Our communal task is to re-create ourselves as responsible members of the earth community, and selected elements of different religions—Jewish, Christian, Buddhist, indigenous traditions, and others—can help us do that task.

I agree with Rasmussen in seeing both signs of hope and despair in contemporary ecological circumstances, and in recognizing the importance of cultural symbols and stories. In contrast to Rasmussen, however,

52. Rasmussen, *Earth*, 265.
53. Merchant, *Death of Nature*, 166ff.
54. See chapter 2 for a fuller discussion of creation's story.

I am less confident about our ability to re-create ourselves, and more confident in the integrity and power of Christian traditions as a whole.[55] The task, I propose, is not to construct a new faith out of fragments of old ones in order to fulfill an environmental need that somehow stands outside of any faith. The task, rather, is to recognize and correct particular errors of theology and practice that lead us (in our sinfulness) to unfaithfulness to the Triune God of Jesus Christ. Part of that unfaithfulness, indeed, lies in behavior that is contemptuous and wasteful of God's creation. Nonetheless, this unfaithfulness is not sin against a generic earth spirit, but sin against Father, Son, and Holy Spirit. As the General Confession in the *Book of Common Prayer* reads, "Most merciful God, we confess that we have sinned against you in thought, word, and deed, by what we have done, and by what we have left undone. We have not loved you with our whole heart; we have not loved our neighbors as ourselves . . . . "[56] Taking God seriously as creator, sustainer, and redeemer of creation means that all wrongs are wrongs against God, for there is nothing "above," "beyond," or more general than God. Likewise, there are no goods that arise from some other spring than the fountain of God's goodness. The *Book of Common Prayer*'s General Thanksgiving expresses gratitude to God for all things: "for all your goodness and loving-kindness to us and to all whom you have made. We bless you for our creation, preservation, and all the blessings of this life . . . . "[57] The world is an arrow that points always to God, whether in faithfulness or failure.

ESCHATOLOGY WITHOUT JESUS?

While I find Rasmussen's cosmology a problematic approach to Christian ecology, I agree with him that cosmology, or a broad spiritual understanding of the world, is critical to transformation of our environmental practices. That is one reason why eschatology is so critical to the project. For Christians, however, generic cosmology is insufficient. Just as Jesus is

---

55. Stanley Hauerwas notes that Rasmussen seems, curiously, to echo the promethean presumption held by the purveyors of ecological destruction (in the name of progress).

56. Episcopal Church, *Book of Common Prayer*, 360.

57. Ibid., 101.

not a universal symbol of redemption, the eschaton, as envisioned in the Bible and liturgical tradition, is not a universal symbol of hope. While Rasmussen's project includes eschatology, his work sometimes exemplifies a sort of de-Christianization of the Kingdom of God.

Rasmussen, for instance, seeks a vision of healed, renewed creation that is somehow intended or destined, yet not tied to particular religious "dogma." This vision is a source of hope; he, likewise, shares concern for the paralyzing despair of well-intentioned people, and wants to emphasize the hopefulness of environmental activism in the face of overwhelming obstacles.

> As more and more people realize the danger [of environmental disaster], their response will include the denial and paralysis that come when we sense the hovering spirit of death. The antidote to denial and paralysis is the generation of hope . . . Are there resources in Christian faith which generate such hope?[58]

The primary "resource" for hope, of course, is the resurrection, from which "hope against hope" was born.

> Resurrection hope, the hope of new birth and new creation, does not guarantee that we will learn to live attuned to nature's life-giving cycles. It does not guarantee that we will repent, be converted and return to our senses in time. *What resurrection hope does do is generate the energy for life which leads to caring and commitment.*[59]

Several implications of this account are worth noting. First, Rasmussen turns to resurrection as an affective tool for people's reconnection to earth in the midst of alienation. Resurrection functions as inspiration and education of human action rather than as display of God's action. Second, resurrection here is quite generic—the defeat of death by life—rather than a particular historical event. This resurrection story stars the disciples (then) and compassionate earth-friends (now) instead of Jesus of Nazareth and God the Father. Finally, the accent

58. Rasmussen, "Hope and the Environment," 19–20.

59. Ibid., author's emphasis.

on current human activity effectively sidelines the Kingdom itself. Rasmussen avoids any interpretation of Jesus' resurrection as the decisive, historical inauguration of a specific future reality. Rasmussen is so sensitive to the idea that the particularities of Christian faith are either earth-defeating or perniciously exclusive (or both) that he sees the resurrection story as, at best, simply motivational—a biblical version of "the power of positive thinking."

Rasmussen is surely right to recognize the centrality of hope in Christian life and the pervasive menace of despair with regard to ecological improvement. Yet, the resurrection is not, according to most Christian traditions, a generic example of good coming out of bad. And most Christians' hope for renewed creation is not based on the idea that, on the whole, life always remains a possibility. On the contrary, the resurrection of Jesus Christ changed the character of the universe by initiating the Kingdom of Heaven. Christians hope in the renewal of creation because God has promised in Christ that all creation will be healed, liberated, and brought to eternal joy in communion with Father, Son, and Holy Spirit. "Just as one's hope for eternal life in God is a correlate of the reality of the promises of the God of his faith, and does not ultimately rest upon either man's desire nor man's hope—so man's vision of the New Creation is a product of God who is affirmed in faith to be a creator, redeemer, sanctifier."[60]

Rasmussen's reduction of eschatology to encouragement accords with his method of making Christianity instrumental to a non-parochial earth faith. In the process, however, eschatology becomes even more peripheral to his project; any inspirational story, it seems, might serve as well as Jesus on the cross.

ESCHATOLOGY WITHOUT DIVINE SOVEREIGNTY?

While eschatology is secondary to Rasmussen's project, it is central to the ecological work of Catherine Keller. What is most interesting about Keller is that she sees eschatology as both complex and inescapable in Western society. Yet, she rejects any conception of the omnipotent God who

60. Sittler, *Evocations of Grace*, 187.

enacts and reigns over the Kingdom. She argues that a sovereign Father paired with a self-sacrificing Son has led to the dual models in human society of domineering oppressor and defenseless victim. (Sometimes, as in Revelation, the helpless victim gets to play the role of absolute conqueror, but the model itself remains the same.) According to Keller, therefore, this understanding of Father and Son must be repudiated. However, this argument exhibits three fundamental errors: first, the phenomenon of humans assuming God's sovereignty as mandate for their own tyranny over other humans and nonhumans is neither ubiquitous nor inevitable. Keller rightly notes the commonality of this iniquity and its connections to apocalyptic undercurrents, but her own counter-examples indicate that it has never been a universal phenomenon. If we look at Christian history as a history of horrific abuse, we will find plenty of examples. If, on the other hand, we look at Christian history as a history of faithful persistence in charity and justice, we will also find plenty of examples: the record is mixed. It is, perhaps, more remarkable that charity and justice persevere at all, and even occasionally prevail, than it is that goodness so often seems trampled by evil.

Second, Keller seems to repudiate God's sovereignty purely on the basis of its effects. The whole tradition is thereby subjected to a utilitarian calculus, one which is narrowly defined in terms of domination. That is, she focuses almost entirely on how belief in God's omnipotence fosters or inhibits abusive power and oppression, leaving out alternative issues such as the belief's promotion of patience, courage, praise, or other beneficial practices. More to the point, this severe utilitarianism adopts a Feuerbachian perspective in the worst sense; that is, she implies that the divine *ought* to be imagined in a way that supports a particular sociopolitical agenda.[61] She writes, "divine force cannot, after the close of the twentieth century, be imagined as an agency subsiding in an omnipotent transcendence."[62] Imagined by whom? And why not? Keller makes little

---

61. Any theologian may fall into the trap of interpreting the doctrine of God in ways most favorable to his or her own interests; continuous conversation with Scripture and the long tradition presumably provides some protection against this trap. To suppose that divinizing human longing is the goal of theology, however, strikes me as perverse.

62. Catherine Keller, *Apocalypse*, 307.

argument for this claim except that to abstract God's agency "from our own is to refuse responsibility for ourselves."[63] Historically, at least, it is not clear why God's transcendence and sovereignty should be more in doubt now than after the close of the fourth or the sixteenth century. Keller might be correct to imply on post-structuralist philosophical grounds that we cannot ever get beyond our signifying to the thing signified—in this case, God. On the other hand, human limitation in understanding and speaking of God is also a long-running thread in Hebrew and Christian traditions. This is usually a claim about human imperfection and sinfulness, though, not about God's unsuitability. Certainly Christian faith should not be unreasonable in the sense of being arbitrary or illogical. But "a philosophical God, the product of our own metaphysical thinking and the construct of our own wayward wisdom . . . is a far cry indeed from the real God who confronts us in judgment and may confront us therefore also in grace."[64] It is not clear how a theism abstracted from the story of Father, Son, and Spirit can help churches as they struggle to respond to ecological issues.

CHRISTIANITY AS OPPRESSOR?

Finally, Keller's portrayal of "classical" divinity as triumphal power is skewed. Despite her reference to "God of power and might" as a key point in liturgy and her intense attention to apocalyptic texts, God's power is *always* paired with God's compassionate love. Thus Christ's crucifixion is not abusive punishment by a domineering Father, but an incredibly generous act by the loving Son.[65] Jesus is not the "oppressed other" who is tortured and killed by a mighty deity, but is himself God. Part of the aim of trinitarian formulas is to make this point: God is one. God is power *and* kenosis, lordship *through* humble service. The tension inherent in this aspect of Christian faith is undeniable, and sometimes leads to a severing between domination on one side and servanthood on the other, with disastrous—apocalyptic—results. Keller tries to escape

63. Ibid.

64 Gilkey, "God," 101.

65. Kelly S. Johnson helped me articulate this point.

the dualism, but her display of an alternative proposal offers little more than egalitarianism and justice with no clear normative grounding.[66]

While Keller is acutely sensitive to the difficulties arising from belief in God's sovereignty, she seems oblivious to the perils of its eradication. To begin with, many—if not most—of the Christians whose own sufferings might have overwhelmed their faith did not, in fact, succumb to their pain. From the biblical stories up to the present, Christians witness to God's abiding presence in their lives *and* God's lordship over the world. Keller's refusal of God's sovereignty in light of their history rejects their own perspective. It amounts to a sort of coercive consciousness-raising: "What you *ought* to have realized is not that you don't understand God's ways, but that God has failed." Such a move not only contradicts the vast majority of Christian traditions, but also undercuts her own emphasis on subverting oppressive domination for justice's sake, because it dismisses the testimony of the oppressed. For example, Brazilian theologian João Batista Libânio emphasizes the sovereignty of God as the source of hope for the poor.[67] Many other liberation theologians make the same point. These claims must be taken seriously if mainstream, first-world theologians are as committed to ecojustice in our own work as we are in corporate and public policy. We cannot afford a theological "holier-than-thou" stance in either the methodology or content of our work. Keller does not intend to take such a posture, but her dismissal of Gustavo Gutierrez and Victor Westhelle as being too "hopeful" in the future of God's promise, as well as her neglect of African-American theological work, have this effect.[68]

Despite her divergence from "classical" theology, Keller is not ready to jettison Christian traditions entirely; rather, she believes it impossible to jettison the tradition that flows so deeply under Western culture. Here she differs from Rasmussen. Rather than constructing an earth faith collage from pieces of various traditions, Keller proposes a "composting" of Christianity, to "sort through the levels of ecclesiastical and theological

---

66. Volf, Review of *Apocalypse Now and Then.*

67. Libânio, "Hope, Utopia, Resurrection," 279–89.

68. On Gutierrez, see Keller, "Power," 60. On Westhelle, see Keller, *Apocalypse,* 168–72.

and perhaps even scriptural chauvinism that have inspired and sanctified our [northern European] empires . . . . Thus we can distinguish between that which can never be composted, which must be eliminated from our history, and that which can be recycled . . . ."[69] In general, however, Keller does not provide adequate justification for rejecting orthodox descriptions of God's sovereignty and redeeming work, especially given her commitment to the victims of poverty, oppression, and ecological devastation. Moreover, precisely because Keller rejects a sovereign god, she lacks the resources to deal with the kind of dilemma I address—between triumphalism on the one hand, and passive despair on the other. It is the promise of Christ's redemption of the world that enables Christians, especially Christians on the "underside" of history, to persist in eco-justice work. Without that promise, Christian activists are, at best, constructing temporary defenses against environmental catastrophe.

ESCHATOLOGY WITHOUT RESURRECTION?

The work of Rosemary Radford Ruether offers a different set of insights and difficulties. Well in advance of many other theologians, Ruether began analyzing the connections between Christian doctrine, Christian practices, sexism, and ecological damage in *Sexism and God Talk* (1983).[70] Her fullest exposition of this subject is *Gaia and God: An Ecofeminist Theology of Earth Healing* (1992).[71] Eschatology is not the focus of this work; however, it figures significantly in the moves she makes away from "classical" Western Christianity. In *Gaia and God*, Ruether tries to construct, from the roots of Christianity, a revised ecological spirituality and ethic. She claims, "there is no ready-made ecological spirituality and ethic in past traditions, because the ecological crisis is new to human experience."[72] By this Ruether does not mean that only moderns have committed ecological harm, but that only recently (after Hiroshima and Nagasaki) have humans realized their potential for global destruction.

69. Keller, "Composting," 166.
70. Ruether, *Sexism and God-Talk*.
71. Ruether, *Gaia and God*.
72. Ibid., 206.

"The radical nature of this new face of ecological devastation means that all past human traditions are inadequate in the face of it."[73] The novelty and urgency of the situation warrant her method of revising classical Christianity to construct a new spirituality. She "sift[s] through the legacy of the Christian and Western cultural heritage to find usable ideas that might nourish a healed relation to each other and to the earth."[74] The goal of her quest is, first, repentance for past and present domination by elite men over non-elite men, women, and non-human nature; and, second, a new path of equitable relationships between God and creation, among humans, and between humans and nonhumans.

On the positive side, Ruether finds "two lines of biblical thought and Christian traditions that have reclaimable resources for an ecological spirituality and practice: the covenantal tradition and the sacramental tradition."[75] In the covenantal tradition, humans and the land are bound together by covenant with God. "The gift of the land is not a possession that can be held apart from relation to God."[76] The fruitfulness of the land requires Israel's obedience to God, and the people's treatment of the natural world is governed by rather strict legislation. In the sacramental tradition, which is more properly called the cosmological tradition, "the cosmos becomes the mediating context of all theological definition and spiritual experience."[77] The Logos-Christ is that which creates, sustains, and reconciles the universe. What is important, here, for Ruether is that in Logos theology, God interpenetrates the world and suffuses it with goodness. God is not an aloof ruler, but an indwelling spirit. So the covenantal tradition "shapes our relationship to nature and each other in terms of law and ethical responsibility." The cosmological tradition "ecstatically experiences the divine bodying forth in the cosmos, and beckons us into communion."[78]

73. Ibid.
74. Ibid., 1.
75. Ibid., 205.
76. Ibid., 211.
77. Ibid., 229.
78. Ibid., 9.

Ruether then moves into physics to note that at the subatomic level, the world appears not as energy and particles, but as a web of relationships. "Matter is energy moving in defined patterns of relationality."[79] But these relationships are also the matrix of all the interconnections of the whole universe. "Thus what we have traditionally called 'God,' the 'mind,' or rational pattern holding all things together, and what we have called 'matter' . . . come together . . . ."[80] God is the whole of the galaxies and the infinitude of miniscule parts. Moreover, because humans are the only self-conscious creatures, we are the mediators between the micro and macro worlds, the site where the universe "becomes conscious of itself."[81] Human consciousness is contradictory: it bears the capacity to roam through space and time, but is completely dependent upon a material existence that has a beginning and an end. (Ruether completely rejects any possibility of afterlife, except in terms of biological composting.) Therefore, an ecological spirituality should be grounded on the knowledge that selves are transient, that all living things are interdependent, and that lives all carry a certain "Thou-ness" in relationship to one another.

How does eschatology figure into this picture? Traditional eschatology, for Ruether, is irreparably flawed. She believes that current theology must accord with—and be limited by—current scientific understandings of physics and biology. Therefore, any idea of individual immortality or resurrection simply denies the reality of life. Just as trees, beavers, and crickets die, humans also die, decompose into elements, and become food or habitat for the next generation of organisms. Consciousness ends when the body dies, although some aspect of individual consciousness may be reabsorbed by the great Matrix. Moreover, she rejects the Christian hope for a final, cosmic consummation for several reasons. Historically, its pedigree is faulty, since it was not part of Hebraic thought, but "imported" from Greek and Persian philosophy. The effects of historical eschatology, moreover, are disastrous. A linear view of history to a final, salvific end point

79. Ibid., 248.
80. Ibid.
81. Ibid., 249.

where goodness has overcome evil yields two possibilities. "Either the end point occurs outside of history altogether and so fails to provide a point of reference for historical hope," or "this final era of salvation is identified with a particular social revolution." In the first case, people suffer a sort of nihilist paralysis; in the second, imperialist oppression and violence. And the "realist" objection applies here, too: eschatological hope, in the classical sense, "contradicts the possibilities of historical existence."[82] It seems, then, that traditional Christian eschatology is rejected *in toto*. Neither the last judgment, heaven, hell, the resurrection, nor the Kingdom of God—as classically understood—can contribute to the new ecological consciousness. It is important to understand, however, that Ruether does not regard her work as jettisoning all hope for the future.

What does Ruether suggest as an alternative scenario? In *Sexism and God-Talk*, Ruether advocates a "jubilee" model of conversion and change. A livable society requires certain ingredients that are both rooted in nature and in the subject of biblical hopes: justice, peace, balanced relationships between creatures, sustainable use of earth, and so forth.[83]

> But human sinfulness creates a drift away from this intended state of peace and justice. Some people's land is expropriated by others. Some people are sold into bondage. Nature is overworked and polluted. So, on a periodic basis (every fifty years), there must be a revolutionary conversion . . . Humanity and nature recover their just balance.

This is not, Ruether emphasizes, a "once-and-for-all" solution, but a "historical project that has to be undertaken again and again in changing circumstances."[84] The new version of a just world will differ from previous versions, although the basic ingredients remain constant. What Ruether does not say explicitly, but what is certainly present in this idea, is that God's activity is more like a vision of the good than an enactment of that good. God/ess sees the way life on earth should work, and participates in human and natural endeavors to correct imbalances, but does not—and

82. Ruether, *Sexism*, 253.
83. Ibid., 254.
84. Ibid., 255.

will not—transcend the limits of biophysical life to fulfill that vision. Individual persons do not survive death, or become resurrected at some final moment. But our lives can still have meaning, in a non-egoistic sense, as our achievements and failures are "assimilated into the fabric of being, and carried forward into new possibilities."[85] Finally, Ruether concludes with this expression of faith and hope:

> It is in the hands of Holy Wisdom to forge out of our finite struggle truth and being for everlasting life. Our agnosticism about what this means is then the expression of our faith, our trust that Holy Wisdom will give transcendent meaning to our work, which is bounded by space and time.[86]

By the time of *Gaia and God*, Ruether's thinking has changed somewhat. The references to the Jubilee year have diminished in favor of the covenantal and sacramental traditions, defined in very general terms. There is more emphasis on the Teilhardian idea of humans as the consciousness of the universe. Her alternative account to traditional eschatology still vigorously rejects a final consummation and individual life after death. Yet again, Ruether thinks that humans are not completely bound by individual finitude, for they have "impulses toward loving care" that somehow exceed the norms of natural life. "These human thinking and caring impulses . . . point to an aspect of the source of life that is also an impulse to consciousness and increased kindness that is still imperfectly realized. We humans are the evolutionary growing edge of this imperfectly realized impulse to consciousness and kindness."[87] This seems to indicate a slow and irregular moral progress over time, although such progress seems to contradict the historical evidence as Ruether herself has described it. Further, elsewhere she rejects liberal progressivism, Marxist utopianism, and Teilhard's confidence in evolutionary progress. [88] So it is not clear how Ruether sees the world proceeding (or not). Given the

---

85. Ibid., 258.
86. Ibid.
87. Ruether, *Gaia and God*, 31.
88. Ibid., 245; *Sexism and God-Talk*, 252.

indebtedness of *Gaia and God* to current astrophysics, we must suppose that she imagines the world ending, eventually, in solar extinction. The human calling is to sustain equitable forms of living and redress wrongs as much as we can, for the extent of our mortal lives. "Then, like bread tossed on the water, we can be confident that our creative work will be nourishing to the community of life, even as we relinquish our small self back into the great Self."[89]

In essence, then, Rosemary Radford Ruether has found (classical) eschatology to be completely antithetical to feminist ecological theology, and has abandoned it. Is such a drastic move necessary? Ruether reads Christian history through an overly negative lens. As this chapter demonstrates, "escape from the body" is not the totality of Christian eschatology, nor is it even the dominant idea in many eras. Moreover, Ruether over-identifies Christianity and Western culture. At the very beginning of *Gaia and God*, she describes Western culture as "enshrined in part in Christianity," and of Christianity as "a major expression" of classical Western culture. "Western Christian tradition is the major culture and system of domination that has pressed humans and the earth into the crises of ecological unsustainability, poverty, and militarism we now experience." Christian cosmologies, doctrines, and practices certainly have shaped Western culture to a large extent, but they have not been the only influences. Great historical shifts are always more complex than a single causal explanation; specifically, the influence of theological—or philosophical—concepts on human action is never straightforward or direct. So to argue that Christian patriarchy produces ecological devastation overstates the case.

Further, Ruether's penchant for identifying patriarchy as the primary human sin understates the variety of human sins and the equal distribution of sinfulness across human divisions: gender, race, geography, class, culture, and sexuality. Systems of oppression often foster different patterns of sinfulness among the oppressors and the oppressed (pride versus self-abasement, for example), but no one is without sin (1 John 1:8–10). Therefore, it is impor-

---

89. Ruether, *Gaia and God*, 253.

tant to see how particular historical oppressions exacerbate—or alleviate—contemporary ecological problems, without excessive generalization.

Ruether attempts to retain an earthly future hope without the telos of the eschaton, because she believes classical eschatology reifies male fears of death. She holds up the vision of Isaiah 65 and Jesus' descriptions of the Kingdom as defining that hope. Both sources of hope, however, become reduced to examples of a universal human hope rather than the specific promises of the Triune God. Jesus comes across less as God incarnate than as a sage who shared and preached a common understanding of the just society. The resurrection, therefore, is not a world-changing event, or even a divine revelation, but a story feeding unrealistic fantasies of immortality. As Peter Phan notes, Ruether's position of "honest agnosticism" about personal afterlife and her argument that "our images of life after death . . . are not revealed knowledge, but projections of our wishes and hopes . . . does not take seriously what has been revealed to us in the resurrection of Christ."[90]

Ruether does not want to base future hope merely on the satisfaction of human desires; she rightly points to the danger of this sort of anthropocentricism. Instead, she sketches the Great Matrix as the basis and the substance of such hope. The Matrix is the web of all inter-relationships in earthly life, from subatomic particles to animals to persons to planets and stars. The Matrix, then, is not a being removed from material existence (as she describes the classical Christian God), but intimately involved in sustaining all life. Moreover, the Matrix is in some way benevolent, or at least life-promoting. "The human capacity for ethical reason is not rootless in the universe, but expresses this deeper source of life "beyond" the biological . . . . To believe in divine being means to believe that [consciousness and altruistic care] are rooted in and respond to the life power from which the universe itself arises."[91]

Ruether's goal is the conversion of Christians to an ecological spirituality and ethic. She deliberately speaks from "within" Western Christian culture, while acknowledging that other religions and cultures must be

90. Peter Phan, "Woman and Last Things," 222.
91. Ruether, *Gaia and God*, 5.

analyzed and reinterpreted by their own adherents. She recognizes the power of one's own tradition, for she writes, "the vast majority of the more than one billion Christians of the world can be lured into an ecological consciousness only if they see that it grows in some ways from the soil in which they are planted."[92] However, I contend that Christians who see themselves as claimed by the biblical story, including such "unrealities" as Jesus' resurrection and the final consummation, are unlikely to recognize the Triune God in the "Great Matrix." Ruether's constructive project will strike many as an alien plant to Christian soil. What (orthodox?) Christians need is not an alternative source of being, but an understanding that the God they have always worshipped will redeem the whole of the universe and, therefore, commands their living witness to that fact. This book is directed to those Christians, typically but not always affluent members of mainstream churches, who are both convinced by the biblical mandate to care for God's creation and uncertain how to do so, or how their efforts can possibly have significant effects.

THE ARGUMENT OF THIS BOOK

Rasmussen, Keller, and Ruether, in their different ways, all see Christianity as requiring significant re-construction or sifting for its contributions to another kind of faith—an earth-friendly faith. They place Christianity somewhere in the middle between the cause of and the solution to ecological destruction. Apologetic writers, on the other hand, place Christianity clearly on the side of the solution. Apologists see Christianity as overgrown by weeds that obscure and choke its ecological guidance. Like me, they emphasize earth-honoring strands within Christian traditions—most often creation, but sometimes incarnation and redemption as well.[93] I have three quarrels with the bulk of this work. First, it sometimes fears misguided religious responses to ecologi-

---

92. Ibid., 207.

93. Andrew Linzey, for instance, works primarily from the doctrine of incarnation, viewing in Christ's self-sacrificial love a model for humans' relationship with the earth. See Linzey, *Animal Theology*.

cal damage more than the damage itself.[94] The anxiety about pantheism, nature-worship, or other sorts of paganism overshadows the concern about creation. But why, in a culture as nature-despising as our own, should nature-worship be of such concern? It is almost as if we hesitate to feed the starving children in Afghanistan lest we make them fat.[95] (Thus I refer to this book as "*boldly* apologetic.") Second, most apologetic works focus on the relationship between humans and nonhumans rather than the relationship between all of God's creation and God. So writers debate whether humans should function as nature's lords, stewards, priests, or partners, rather than addressing the larger story of God's generous love for all. The story of humans' proper place (or proper places) in creation can only be told in the context of the larger story of God's creation, sustaining, and redemption of the whole universe. Third, many apologetic Christian ecologists describe well how individual Christian virtues pertain to ecological living—humility, hospitality, and hope, for instance. At the same time, they overlook the power and responsibility of the church to exemplify, as the eschatological body of Christ, God's harmonious relationship with creation.[96] Ecclesiology is treated scantily, if at all; the church often seems suborned to the generic category of community. I will argue, in contrast, that the churches are the addressee of Christ's call to witness—in particular, the call to witness to what Keller terms "the green eschaton."

In relation to "earth faith" approaches and other Christian calls for improved care for creation, this book stands on the apologetic end of the spectrum, but it focuses on the larger story than that of humans'

94. For a clear example of this attitude, see the Assemblies of God statement on the environment, http://ag.org/top/beliefs/contemporary_issues/issues_02_environment.cfm (accessed March 15, 2005).

95. The Old Testament warns against pagan nature-worship; and it was a real threat to the early church. And it may remain a factor in some segments of our population. It is not nearly such a threat, however, within the church as other forms of idolatry, especially Mammon. The number of our congregants who value material prosperity over faithfulness far exceeds those who value non-human species over faithfulness.

96. Both Sean McDonaugh and Steven Bouma-Prediger lean in this direction, although their work is generally admirable. McDonaugh, *Passion for the Earth*; Bouma-Prediger, *Beauty of the Earth*.

role vis-à-vis nonhumans. It focuses, instead, on the church as witness-ing body to the Kingdom of Heaven, which, as promised by God, encompasses the redemption of all God's creation. Throughout this work, I endeavor to pay explicit attention to the theologies of those groups that suffer the worst effects of environmental destruction—liberation theolo-gies of Latin Americans and African-Americans. My primary goal is not to make the church a participant in the "environmental movement," but to make the church more faithful by including the eschatological import of creation in witness to Jesus Christ and the Kingdom of God.

Chapter 2 will explore in depth the doctrine of creation and its implications for eschatological ethics. In order to understand how humans should live on the earth, it is critical to understand what the earth *is*, on Christian terms. The doctrine of creation is not the same as the "common creation story" that some environmentalists hope can transcend differences in culture or belief. Rather, it is a very particular story of the Triune God's free, loving, and ongoing activity in fulfillment of the divine plan. Creation, therefore, is inherently eschatological and christological: created by God through Jesus Christ to be sustained in the Spirit and ultimately brought back to glorious unity with the Trinity. In the meantime it reflects God's grace, but is also distorted by sin and corruption. It is not a single, unitary thing, but a diverse and pluriform array of creatures and environments, all held in relation to each other and to God through Christ.

What difference does this make for ethics? Chapter 3 articulates a view of Christian ethics as communal witness to God's ongoing gra-cious activity in creation and to God's promise of creation's fulfillment in the eschaton. Christians are not commanded to "fix" the world, but to witness to God's redeeming presence and activity already in the world. In particular, Christians witness to the Kingdom of Heaven that was inaugurated by the life, death, and resurrection of Jesus and will be con-summated in his coming again. Christians' knowledge of the kingdom is neither dispassionate cognition nor laboratory science. Rather, it is only in faithful living, in following Jesus, that Christians get a glimpse of the possibilities of the kingdom. Peace, justice/liberation/reconciliation, abundance, righteousness, and communion with God characterize the promised Kingdom. Christian witness, therefore, consists of the community of faith demonstrating the possibilities of all these things.

Witness work is not building the Kingdom, or transforming the world, but living in testimony to Christ's transformation of the world. It is costly, laborious, and even dangerous work, for it reveals by contrast the powers and principalities of the "anti-Reign," to use Jon Sobrino's term. Witness work is done in obedience to God, and thus performed according to faithfulness instead of means-end effectiveness. This is not an invitation to stupidity, but a reminder that Christian prudence is sometimes the opposite of secular prudence (which usually equates with means-end thinking). Prudence for Christians is the practical reasoning toward the action that best testifies to God's love for the world and to God's promises of the world's salvation, even if the short-term prospect is dismal failure. Christian witness, therefore, is hopeful, joyful, open to creative alternatives, and communal.

How the church embodies this kind of witness is the subject of chapter 4. The argument begins, first, that Christian ethics is inescapably ecclesiological; the addressee of ethical mandates is the church, the body of Christ.[97] This claim is anthropological and ontological as well as political. A living community of faithful people is critical to the embodiment of Christian discipleship. Moreover, humans and all members of non-human creation are essentially social, as they reflect the innate sociality of their creator, the Triune God. It is the church, therefore, that posits signs of God's Kingdom in a sinful world. Once again, we should remember that the church does not build or usher in the Kingdom; for the church, like the rest of creation, is a gift that honors its creator, not a partner of the creator. The task for the church is to be in the world as a living image of submission to the lordship of Christ.[98]

In order for the church to witness it must first be visible, not having "disappeared" into innocuous assimilation with secular culture. It must maintain a balance between verbal and nonverbal proclamation—in other words, between formal liturgy and "outreach" work, so that it never

97. Hays, *Moral Vision*, 196.
98. Cone, *Black Theology of Liberation*, 132.

43

loses its eucharistic identity as the "civic assembly of the eschatological city," to use David Yeago's words.[99] Because it witnesses not to any generic utopia, but to the consummation of *this* creation through Jesus Christ, the church's relations with its earthly community should manifest Christian peace, abundance, justice, liberation, righteousness, and communion with God. These eschatological relations include such aspects as a minimum of violence toward the parish's local land, animals, and ecosystems; and a joyful frugality and ecojustice in the consumption of natural resources, *especially* in the bread and wine of Holy Communion. These signs of the Kingdom require courage, creativity, self-questioning, patience, and perseverance. Thus are we brought back to the Christian virtues, the gift and development of which enable the church to live into its witness. Patience is second only to faith in its centrality to ecological discipleship. Even the word "witness," in one of its meanings, signifies watching—and watching is an activity of patience. Christian patience consists of the endurance of adversity, the hopeful waiting for the coming of Christ, and the receptive delight in the particularity of something, a kind of appreciation of the created world for what it is, rather than for what we can make it become. Only with such patience can the church enact the body of Christ in its life on earth.

Throughout the book, then, I show how the church's faithfulness requires ecological discipleship on the church's own terms. The point is not to make the church a participant in the "environmental movement," but to make the church more faithful by including the eschatological import of creation in its performance of worship—worship that consists not merely of weekly services, but of a "way" of life that praises and witnesses to Father, Son, and Holy Spirit.

99. Yeago, "Messiah's People," 150.

CHAPTER 2

GOD'S ESCHATOLOGICAL CREATION

CONTEMPORARY THEOLOGIANS often claim that creation is ineluctably connected with eschatology. The reaffirmation, however, of creation's eschatological character is quite recent—occurring, with a few exceptions, within the last twenty years. It has taken some time for the early twentieth-century renewed interest in eschatology to unite with the later-twentieth-century interest in creation theology. Even now, only a handful of theologians have explicitly addressed the connection between God's creation and the Kingdom of God. The work of Ruether, Rasmussen, and Keller was examined in chapter 1 in this regard. Ruether and Keller want to reject certain basic assertions of Christian eschatology, while Rasmussen wants eschatology without christology. In contrast, I argue in this chapter that the created universe is fundamentally eschatological and christological, and that the eschaton is part and parcel of God's creative activity. Neither creation stories nor end-of-the-world stories are unique to Christianity; indeed Brian Swimme has proposed a "common creation story" to inspire interfaith ecological efforts. For Christians, however, the story of creation and consummation are not generic, but narratives of God's activity for, and in, the world through Jesus Christ. Exploring the density and range of the creation-eschatology connection is crucial, therefore, for understanding how God's creating and sustaining presence relates to God's intentions for the material world.

The chapter begins with an assessment of the "common creation story" and several cautions about our discourse regarding nature and creation. It then addresses the doctrine of creation in two sections: first, the activity of God, and second, the results of God's activity. The cumulative

45

effect of these analyses should be an image of the created universe as a finite, beloved, good-yet-flawed, tremendously varied universe of particular beings, forces, and relationships, all being carried toward their fulfillment in God's purposes.

COMMON CREATION STORY?

⅋ Contrary to the claims of Brian Swimme and Thomas Berry, there is no universal story of creation. Swimme and others have offered a "common creation story" drawn, in part, from popular versions of contemporary science, in part from biblical elements, and in part from current ecological motifs. So it begins,

> Originating power brought forth a universe. All the energy that would ever exist in the entire course of time erupted as a single quantum—a singular gift of existence. If in the future, stars would blaze and lizards would blink in their light, these actions would be powered by the same numinous energy that flared forth at the dawn of time. There was no place in the universe that was separate from the originating power of the universe. Each thing of the universe had its very roots in this realm. Even space-time itself was a tossing, churning, foaming out of the originating reality, instant by instant. Each of the sextillion particles that foamed into existence had its root in this quantum vacuum, this originating reality.[1]

The term "common" points to both the universal origin of all things (the big bang) and the universal appeal of such a story to religious and nonreligious alike. Swimme writes,

> The creation story unfurling within the scientific enterprise provides the fundamental context, the fundamental arena of meaning, for all the peoples of the Earth. For the first time in human history, we can agree on the basic story of the galaxies, the stars, the planets, minerals, life forms, and human cultures. This story does not diminish the spiritual traditions of the classical or tribal

1. Swimme and Berry, *Universe Story*, 17.

periods of human history. Rather, the story provides the proper setting for the teachings of all traditions, showing the true magnitude of their central truths.[2]

The intentions behind this project are benevolent. Berry and Swimme are targeting people who feel caught between science and religion, or who feel alienated from both. They want to provide such people with a positive perspective from which to envision their place in a harmonious universe, and hence to modify their lives in earth-friendly ways. Moreover, Berry and Swimme are aware of the importance of story in human identity: who we are depends on the stories we, and others, tell. While the "common creation story" project utterly fails, its failings are instructive for how Christians should approach the diverse understandings of creation, even within their own tradition.

First, the common creation story is not universal even among the experts who supposedly discovered it. While the broad outlines of space history and evolutionary biology are generally accepted among Western scientists, the particular sequence, timeframes, and connection of events are subjects of disagreement and uncertainty.[3] Second, to call this version of contemporary science a story that "we can all agree on" immediately begs the question: What "we"? Are we that "we"? It is neither a story that would appeal to many Protestant evangelicals, nor perhaps to conservative Muslims, Buddhists, Hindus, nor goddess followers, to name a few million. Even people who would accept it as a sort of scientific creation story might bridle at the notion that it "embraces" their religious faith.[4] The overlap or consonance of creation stories—like any fundamental narrative—cannot be imposed by colonial decree; rather, it must be discovered in the mutual telling of those stories. Moreover, this "common creation story" is predicated on the abstraction and objectivity

---

2. Swimme, *Green Dragon*, 38.

3. A recent issue of *Scientific American* (May 2004, vol. 290:5) carried the headline: "The Big Bang May Not Have Been the Beginning."

4. "The Great Story is a way of telling the history of everyone and everything that honors and embraces all religious traditions and creation stories," www.thegreatstory.org/what_is.html (accessed February 21, 2005).

of Western scientific language, as if it is somehow arching above all the local stories that it can "embrace." Scientists (the cytogeneticist Barbara McClintock, for instance), who point out how particular cultural assumptions have governed and shaped Western scientific endeavors, have refuted such objectivity.[5] Swimme seems to think the universal creation story can merge scientific and religious discourse; what actually occurs, though, is a subordination of religion to science—scarcely a new or desirable phenomenon in Western culture. Furthermore, as Reinhard Hütter points out, the Christian creation story is less about the "nature" of the reality (as Western science sees it) than about "the One who has brought this reality into being and is present in this reality as creative agent."[6]

UNIVERSAL NATURE?

❧ A universal story of creation is as unlikely as a universal understanding of "nature." The terms "nature" and the "natural" slip and slide over meanings, depending upon political aims and discursive contexts. Raymond Williams has pointed out that the earliest meaning of "nature" in ancient Greek was probably "the inherent and essential quality of any particular thing."[7] This gradually became generalized, in medieval Christian orthodoxy, to mean the nature of all things, the "essential constitution of the world."[8] Nature pointed to a greater abstraction, a universal principle around which the world's innumerable objects and processes could be organized. We still find this usage in statements about the nature's power of destruction or nature's tendency toward greater complexity. Most often in environmental discourse, however, "nature" refers to non-human biophysical reality. Nature is the land, sea, air, and outer space; it is the animals, plants, minerals, climate, earthquakes, hurricanes, and dirt. Nature, in this parlance, does not usually refer to human beings, cities, human constructions, ideas, or artistic endeavors. So Western

5. Keller, *Feeling for the Organism.*
6. Hütter, "*Creatio ex nihilo,*" 91.
7. Raymond Williams, *Problems,* 68.
8. Ibid.

48

environmentalists want to conserve or protect "nature" from the ravages of human beings. Obviously, however, this conception of nature is under constant dispute. Are the Dakota Badlands a United States national park and treasury of prehistoric fossils, or are they the sacred lands of Sioux ancestors? Are the Amazonian forests a wilderness to be protected from human incursion or are they the home of local Indians? Is the patch of ground beside a housing project a valuable bit of nature to be nurtured and gardened, or is it a wasted piece of "real estate?" In all such conflicts, the meaning of nature never stands alone, but stems from complex intersections of history, geography, social location, religion, race, class, and gender. We might ask, "tell me what you see in the land, and I can tell who you are."

The conflicts over nature, therefore, are not apolitical, but always involve clashes of interest and power.[9] For instance, the numerous legal battles between the U.S. government and Native American tribes involve different understandings of land, divine presence, and property ownership, as well as sovereignty and territory. In another arena, white feminists of the 1970s sought to reclaim the positive aspects of women's "connection" with nature—thus highlighting women's "natural" tendencies toward care, nurture, and creativity. At the same time, however, black women were struggling against longstanding stereotypes of themselves as more "natural"—i.e., beastly, sexual, emotional—than either black men or white people.[10] Any attempt to target nature or the natural must be open to scrutiny of its socio-political presuppositions.

What does all this have to do with Christian theology? "Nature," after all, in the sense of non-human biophysical reality, is not a specifically theological or biblical category.[11] Christians generally speak about "creation" instead. Nevertheless, the histories of Christian attitudes toward God's creation and secular attitudes toward nature cannot be completely separated. While Christianity is not to "blame" for the current ecological

9. Soper, *What is Nature?*, chapter 1.

10. Thistlethwaite, *Sex, Race, and God*, 58–59.

11 Paulos Mar Gregorios notes that the concept of 'nature' in the sense of non-human, self-existent reality does not occur in the New Testament or Old Testament. "New Testament Foundations," 40.

degradation, Christian doctrines certainly factored in the development of modern science in the sixteenth century, and continue to be wielded as justification for anti-environmental behavior. Moreover, the traffic flows both ways: shifts in Christian doctrine and practices are always located in or against secular events and cultural trends. So it is never as easy as saying, "Whatever non-Christians think, Christians believe that God's creation is valuable and should be cared for by humans."

A second reason for Christian theologians to attend to the discourse of "nature" is that "creation," too, is a concept with political effects. The confession that "God the Father, God the Son, and God the Holy Spirit created all that is" bears substantive content, as we shall explore further. Its interpretation and implementation, however, inevitably produce differences in understanding, in part because they are subject to all the earthly influences of race, class, history, geography, and other factors. So the medieval bestiaries' presentation of animals as moral instructors is quite different from the Puritan dread of ungodly wilderness, and both of these differ widely from the characterization of nature in Athanasius' *Life of Anthony* as a populace of "junior monks" that can be taught Christian obedience.[12] So we need to attend to the specific context of different manifestations of creation doctrine, as well as their effects on local human and non-human communities. As Kate Soper writes, the recurring motif of Eden as a source of purity "has been a component of all forms of racism, tribalism, and national identity."[13] Christian dogma is not responsible for current environmental destruction and ecological injustice, but neither can it dispute its involvement in these matters, for dogma arises from, and points toward, human life under God on earth.

About "creation" we must also ask not only "whose creation" and "what Christians," but also "which part of creation?" Any strong dichotomy between A and B works to magnify the differences between A and B and blur the differences within A and within B. Thus, the modernist dichotomy between humans and nature has perpetuated and masked unjust distribution of natural resources and environmental damage, by

12. Yordy, "Eco-Critical Reading."

13. Soper, *What is Nature?*, 18.

abstracting both humans and nonhumans from their particular places and lives. A 1989 special edition of *Scientific American*, entitled "Managing Planet Earth," said,

> It is as a global species that we are transforming the planet. It is only as a global species—pooling our knowledge, coordinating our actions and sharing what the planet has to offer—that we may have any prospect for managing the planet's transformation along the pathways of sustainable development. Self-conscious, intelligent management of the earth is one of the great challenges facing humanity as it approaches the twenty-first century.[14]

Here we have the planet as a single, unitary object to be managed by a single, unitary subject—humanity. The enormous, interrelated pluriformity of creatures, plants, soils, minerals, lunches, dwellings, burial rites, gods, tools, and so forth becomes two great abstractions on either side of a divider: nature vs. humans. This perspective masks important realities on both sides of the divide, which itself obscures the deep and intricate interactions between cultures and their environments.

On the one side, "the environment itself is local; nature diversifies to make niches, enmeshing each locale in its own intricate web."[15] A tree is never a generic "tree," but is always a tree of one particular type or another. A stream never flows in the abstract, but always in this place or that place, around these particular rocks. Certainly humans and nonhumans can appreciate commonalities among trees and streams. But "nature" abstracts from the very specificity inherent to ecology and to ecological trauma. It is, in this sense, a contradiction in terms. As E. O. Wilson says, "Extinction is the most obscure and local of all biological processes. We don't see the last butterfly of its species snatched from the air by a bird or the last orchid of a certain kind killed by the collapse of its supporting tree . . . ."[16] To say, for instance, "natural water resources are endangered" obscures the locative materiality of the local creek, its particular flow

---

14. Quoted in Sachs, "Global Ecology," 18.

15. Lohmann, "Resisting Green Globalism," 159.

16 Quoted in Thomashow, "Toward a Cosmopolitan Bioregionalism," 122.

patterns, chemical threats, water oaks, and spotted salamanders beneath the placeless constructions of "water resources" and the "natural."[17]

On the other side, the language of "nature vs. humans" identifies the environmental culprit as the human species. "Humans" here are also taken as a generic type, especially in discussions of population explosion and its impacts upon planetary ecosystems. In terms of biophysical consumption and production, however, humans are not all equal. The oft-cited statistic is that the twenty percent of the population living in Europe, North America, Oceania, and Japan are using roughly eighty percent of the planet's resources and its garbage depositories. "If the South disappeared tomorrow, the environmental crisis would be still with us, but not if the North disappeared."[18] More specifically, an American child will use, during her lifetime, many times the energy, food, life, and water as an African child. But the image of the human-nature dichotomy obscures these huge disproportionalities.

Now "creation," in contrast to "nature," may transcend the dichotomy between humans and nonhumans. The doctrine of creation at least relativizes differences on earth against the great difference between creature and Creator, as we will see later in this chapter. This very fact, however, often obscures the great particularity of God's material creation: God did not create a single "thing," but a zillion different things, each with its own relationship to other kinds and to its Creator. This diversity and complexity is reflected in the Genesis account of different "days" of creation, and, to an extent, in biblical distinctions between domestic and wild animals. Similarly, Christian attitudes toward "wilderness" have experienced a different sort of history from Christian attitudes toward farms and gardens.[19] Those differences should, I believe, be honored in our efforts to comprehend and improve our ecological discipleship. A tiger has

17. Of course, it is impossible to write about "nature" or "creation" without using some abstracting language. Here I shall try to use "the created order" or "the universe" as a rough synonym for the Greek *ta panta*: "everything that is." In this context the universe might, in fact, be many universes; my point is to try to include all that is, without forgetting the infinite differences between the individuals that comprise "everything."

18. Banuri, "Landscape of Political Conflicts," 50.

19. See, for instance, Bratton, *Christianity*.

more in common with a man than with a stone, and the tiger, stone, and man all share more with each other (I shall argue) than with their God. Disregard of these differences may arise from fears that humans will lose their "special place" in the created order, but it contributes to theological muddles as well as uncharitable behavior toward our fellow creatures.

All of this, of course, is not to reduce the doctrine of creation to its historical settings or to an ecological treatise on biodiversity. Such a move would deny our own historicity and any claim that our own traditions hold on us. The point is simply a caution that our own work is as contextual as anyone else's, and that part of our context includes a legacy of using generalizations about the created world to uphold unjust structures of power and domination. What follows, then, is not a universal narrative in the sense of being claimed (or claimable) by everyone, though it shares elements with other creation stories. Rather, it is the Christian doctrine of creation as it developed out of the Israelites' and early Christians' understanding of God. The story and doctrine laid out here are ecumenical in character, although certain points have been disputed by Christians at various places in various times. Given the desire to be sensitive to the socio-political effects of "creation" just as we are to "nature," I note contentious or perilous aspects of the Christian understanding of creation.

The Christian creation story exhibits a number of differences with the common creation story, some of which are implicit in the description below. The key difference for our purposes is that for Christians, creation is eschatological. It is not a thing to be preserved in stasis, but a universe of things and subjects and processes and relations all on their way to unity in God. Our job, then, is not to preserve the "thing" as if it were an object in lucite, but to honor the plethora of journeys and to witness to their future possibilities in Christ.[20]

One of the primary messages of the Bible, delivered throughout and in many different styles, is that God is the origin and Lord of all. Thus it

---

20. This is not the place for an extensive treatment of the theme of Christian pilgrimage, but it may be fruitful to take seriously the idea that every part, parcel, penguin, and pineapple in creation is on its own pilgrimage to God. It would be important, of course, not to view such pilgrimages as atomistic, but somehow to conceive them as individual, yet interdependent.

is only right to begin a review of Christian doctrine about creation with attention to God's creating activity. The universe, including its materials and energies and inhabitants, is the result of God's activity, but all the power and initiation are God's. And if Christian life—in particular, our earth-oriented life—is our human response to the divine creating, sustaining, and redeeming work, it is helpful to review how that work is understood.

The very fact that God is Creator entails that, for Christians, the created universe is not a human construct, but reality. It is, of course, perceived and experienced and formulated in quite different ways by different creatures in different circumstances (including human creatures), but it is not a human invention. Creation, including us humans, is a God-made "thing." The universe, the planet, and the divine intention precede us. In Wendell Berry's words, "we are living from mystery, from creatures we did not make and powers we cannot comprehend."[21] That being so, "created nonhuman being essentially transcends the circle of the human story." It is bigger—physically, morally, and teleologically—than us humans.[22]

GOD'S CREATING ACTIVITY

❧ What, then, do Christians understand about God's creative action? George Hendry writes that because God's creation is unique, and because it is the precondition of our experience rather than an object of our experience, it can be described only analogically. Christians have employed different models of creation, each with its advantages and disadvantages: procreation, fabrication (making), formation (molding), invention, expression, and emanation.[23] Procreation and emanation overstate the tie between God and the universe, while fabrication, formation, and invention may understate it.[24] Formation implies pre-existing material to be

21. Berry, *What Are People For?*, 152.

22 Lowes, "Up Close and Personal," 135.

23. Hendry, *Theology of Nature*, 148ff.

24. The procreation model contradicts the church's claim that only Jesus Christ is begotten of God; God creates all else. On the other hand, feminists such as Catherine

formed, which limits God's power and the scope of God's sovereignty (see the section on creation *ex nihilo*, below). Expression, therefore, may be the best analogy; God creates through self-expression of the divine love. Nonetheless, the other analogies should not be eliminated; a variety of models has biblical precedent and helps remind us that God's creative activity is not really like anything humans can describe or even imagine.

There has been broad agreement (though not unanimity) on several key elements of the act of creation: God's freedom and graciousness in creating, God's creating out of divine generosity and love, God's creating as trinitarian activity, God having created *ex nihilo* (these last two points were disputed in the early centuries, but rarely now), and God's creating as a continuous, sustaining activity that follows God's ultimate intentions for the created universe.

Creation as a Free Act of God

God is complete in Godself; God does not need this (or any other) universe for completion. God chooses to create out of perfect freedom, so all of creation depends for its existence upon the divine will. This idea has made some theologians nervous because it seems to make the natural order tenuous, as if a momentary fancy or fleeting inattention on God's part would destroy everything instantly.[25] Moreover, "if the divine will is overstressed, then God ends up seeming indifferent to creation—whether or not to create is a trivial decision on God's part."[26] So writers as diverse as Schiller, A. N. Whitehead, and Jonathan Edwards have posited that creation was in some way necessary for God—either an inevitable outgrowth of God's loving nature, or a completion of God's being.[27]

Keller and Rosemary Radford Ruether strongly criticize the fabrication model as presenting the divine as aloof and largely indifferent to created reality. See Keller, "Power Lines," 55–77.

25. John Calvin did in fact believe that only God's constant restraining hand kept the creation from erupting into wholesale chaos, with elephants and other wild beasts charging into cities and attacking humans (Calvin *Commentary on Genesis* vol. I, IX, 2). For Calvin, though, this threat of disorder was a result of the Fall, not of God's freedom.

26. Hendry, *Theology of Nature*, 124.

27. Hendry, *Theology of Nature*, 121; Jonathan Edwards, "A Dissertation," 403–536.

Freedom, however, need not entail caprice, indifference, or unreliability, and desire is not compulsion. God desires and freely commits to a loving, providential and sovereign relationship with his creation, as testified to in Scripture.[28] (One human analogy to this is the free commitment people make to bind themselves in marriage, ordination, parenthood, or any other lifelong obligation.) Richard Hooker has a helpful way of describing the connection between divine love and divine freedom in creation: God freely consents to the limits set to divine action by divine nature.[29]

Moreover, viewing creation as necessary for God poses dangers not only for our understanding of God (by limiting God's freedom, power, and transcendence), but also for our understanding of creation. Affirming God's freedom in creating (or not creating) provides us with a perspective on creation that avoids utter chance on the one hand, and inevitability on the other. The natural order was created according to God's will—whether that will was fulfilled through Big Bangs or little whispers. Had God willed otherwise, the universe might have been entirely different. (The use of the past tense is, of course, a concession to our own finitude, for time and space are elements of the creation, not of the creator.) The creation event, the created universe, and all its creatures are contingent, in being and nature, on the will of God. The universe in its fallen, corrupted state imperfectly fulfils the divine intent, but its being and enduring are according to God. Thus, despite the apparent serendipity of life on earth from a scientific perspective, the universe is not a product of chance, for God's intent is neither random nor capricious. On the other hand, we should not, as some environmentalists do, absolutize the natural order in which we participate, because it all could have been otherwise. This path between chance and inevitability, in turn, leads to a freer eschatology. Our visions of the Kingdom of God can be expanded by the knowledge that God's consummating activity need not adhere to the "laws of nature" as we understand them.

28. Schwöbel writes, "creation therefore has to be seen as an expression of the love of God who remains faithful to what he's created in love—not a temporary attitude adopted by God's will, but a relationship anchored in God's being," (Schwöbel, "God, Creation, and Community," 156).

29 Rowan Williams, "Hooker," 371.

The Ground of God's Creating Activity

This is connected to, but not identical with, the previous point about God's freedom. God freely created and sustains all that is, but the reason remains elusive from a human perspective—it is, in fact, presumptive to speak of this sort of "reason" in regard to the divine. Given the infinite goodness of God, we can rule out that creation is a bad joke, a vehicle for suffering, or a theater of the absurd, although experience sometimes tempts us toward these explanations. For the most part, Christians have followed the emphasis in Scripture on God's self-giving love as a reason for God's creating, sustaining, and redeeming activity. Being is better than non-being, so bringing something good into the world is an act of generosity and charity.

What is important for this discussion, more so than the accepted benevolence of God's creating activity, is that creating was not a one-time splurge of generosity on God's part. Instead, "creation is ever-fresh, a continual grant of existence from the single and holy divine fount of being."[30] God's creation of the universe is part of the constant divine intention that the world be created, sustained, redeemed in Jesus Christ, and ultimately brought to complete fulfillment in the Kingdom of God. The early Christians' astonishing discovery was that the man they knew as Jesus, who announced the coming of the Kingdom, was the same God that created the universe; he was also the same Spirit who sustained the created universe, as well as God's people, throughout their history and across vast geographies. The divine love that generated the universe neither ceased on the seventh day, nor dissipated in the following centuries; and neither is it content to continuously pull creatures out of their self-made quandaries. The love of God will bring the universe and all its bits and pieces and inhabitants and energies to a consummating communion with Godself. Moreover, affirming that God's loving care for creation is ongoing and participatory reminds us that creation is not a static backdrop for human progress, but an innumerable diversity of characters with their own story of life under God's providential plan. At its most basic level, the universe is story, or rather, a storied relationship.

30. Carmody, "Ecological Wisdom," 100.

This is a legitimate point raised by Swimme's "common creation story": "the cosmos itself becomes a narrative, a grand adventure . . . filled with restlessness, suspense, and mystery."[31]

CREATION AS A TRINITARIAN ACTIVITY

Affirming the distinctive roles of Christ and the Spirit in creation—rather than simply identifying the act of creation with God "in general"—follows New Testament and creedal testimony. The most obvious New Testament texts referring creation to Christ are the prologue of John, 1 Corinthians 8:6, Colossians 1:15–20, Philippians 2:6–11, and Hebrews 1:2–4. With variations in phrasing, all of these verses confess "one Lord, Jesus Christ, through whom are all things." (1 Cor 8:6) In Colin Gunton's words,

> under the impact of the resurrection the belief developed early that one whose relation to God was of such a kind was, indeed, the one through whom God made the world . . . However remarkable the claim, whatever the origin of the words used to express it, and however incredible may appear the content, there appears to have developed soon after the death of Jesus a widespread Christian confession to the effect that the one through whom God had acted to save the world was also the agent of its creation.[32]

In accordance with this fundamental element in early Christian belief, the Nicene Creed includes in its second article "We believe in one Lord, Jesus Christ . . . .Through him all things were made." There can be no question that from the very early days of the Church, Christians have attested to the agency or mediation of Christ in the act of creation, and the activity of the Spirit in sustaining and enlivening creation.

Part of what is involved in such a claim, clearly, is that Jesus Christ is truly God, one with the Father. Thus Athanasius, writing about the creation, says,

31. Conradie, "Stewards or Sojourners," 165.
32. Gunton, *Christ and Creation*, 22.

. . . there was necessarily someone with [God], to whom he spoke when making the universe. Who then could it be except his Word, for to whom could one say God speaks except to his own Word? And who was with him when he was making all created being except his wisdom, who says: 'When he made heaven and earth I was with him'? . . . Being with him as wisdom, and as Word seeing the Father, he created the universe, formed it and ordered it; and being the power of the Father, he gave all things the strength to come into existence . . .[33]

Basil, likewise, uses the description of Christ as the "power of God" to emphasize the oneness of the work of Father and Son even as he points out the distinctive activity of the Son:

The Lord says, "All mine are thine," as if He were submitting His lordship over creation to the Father, but He also adds "thine are mine," to show that the creating command came from the Father to Him. The Son did not need help to accomplish His work . . . Rather, the Word was full of His Father's grace; He shines forth from the Father, and accomplishes everything according to His Parent's plan. He is not different in essence, nor is He different in power from the Father, and if their power is equal, then their works are the same. Christ is the power of God, and the wisdom of God.[34]

If all things were made through the Son, then the Son is fully divine; at the same time, if the Son is indeed the Lord, then he must be co-eternal with the Father and the one through whom the universe was created. The two assertions are interconnected and mutually supportive. Thus, for instance, Calvin "insists that proper consideration of the power which Christ exercises as Logos, or instrumental Author of creation, actually proves "his eternity, his true essence and his divinity."[35] Retaining distinctions among the divine persons in the economic activity of creation ensures, moreover, that the Triune God confessed by Christians is the

---

33. Athanasius (Thompson, 129–31) *Contra Gentes.*

34. Basil the Great (Anderson, 39) *On the Holy Spirit.*

35. Calvin, *Institutes*, 56.

same God who created the world, the same God whose creative act is told in the Hebrew Scriptures.

This last point leads to Catherine LaCugna's argument that failure to recognize real distinctions among the divine persons either in the immanent Trinity or in the economic Trinity hardens the breach between *oikonomia* and *theologia*. The Trinity is not something we can assert in one aspect of God's relationship to us and downplay (or ignore) in another. "Because the personae of the Godhead represent real hypostases in God, it is legitimate and even desirable to refer to God's relations with the world according to the differentiated personae."[36] Recognizing "the uniqueness of the missions of Son and Spirit" in creation helps retain the integrity of our belief in the trinitarian God, and supports our conviction that that Triune God is the One who creates, redeems, and sustains the universe.[37]

That God is One entails a strong, integral connection between creation and salvation, a connection that is trinitarian—and particularly christological—in nature. This connection is, in a sense, the economic basis of trinitarian doctrine. Joseph Sittler says of Christ, paraphrasing 1 Colossians 15–20: "He comes to all things, not as a stranger, for he is the first-born of all creation, and in him all things were created. He is not only the matrix and prius of all things: he is the intention, the fullness, and the integrity of all things: for all things were created through him and for him."[38] But this christological connection does not contradict the heritage of Israel's life under God. For as Gunton points out, the Old Testament conceptions of salvation "are framed within beliefs about the creation of the world."[39] And as Israel's experience of salvation (through exodus and covenant) revealed God's hand as creator of the universe, so the Christian experience of salvation through the risen Christ revealed Christ as the agent of that creation. Jürgen Moltmann puts it even more

36. Wyatt, *Jesus Christ*, 57.

37. LaCugna, *God for Us*, 99.

38. Sittler, "Called to Unity," 177–78.

39. Gunton, *Christ and Creation*, 20. Note that LaCugna, Sittler, and Gunton do not agree about everything in trinitarian doctrine; these agreements they do have, then, are all the more important.

strongly, saying that "if Christ is the ground of salvation for the whole creation, for sinful men and women, and for "enslaved" non-human creatures, he is then also the ground or the existence of the whole creation, human beings and nature alike."[40] Thus, many New Testament texts about creation speak of it in past, present, and even future tense, linking God's "original" creation of the world through Christ with God's continuing sustenance of the world—again through Christ—with Christ reclaiming creation in the future. This connection becomes explicit in wisdom christology, where "the mediation of glory and creation are one and the same."[41] Jesus Christ, understood as the wisdom of God, is the one through whom God creates and orders the world, and through whom God will one day glorify the world.

The connection between creation, incarnation, and redemption through Christ is a vital strand in the Christian tradition, and has been expressed by theologians as diverse as Pope Leo I, Joseph Sittler, and Bruno Forte. Speaking of Mary, Leo writes:

> Only then, with Wisdom building a house for herself would the Word become flesh within her inviolate womb. Then, too, the Creator of all ages would be born in time and the nature of God would join with the nature of a slave in the unity of one person. The one through whom the world was created would himself be brought forth in the midst of creation.[42]

Creation, then, is christological and eschatological. It is *this* world that was created by Father, Son, and Holy Spirit, and *this* world that is redeemed by Jesus' death and resurrection, and *this* world that will be taken up into final consummation with the Trinity.

A final reason for asserting the that trinitarian roles in the act of creation is epistemological is that, only with a strong doctrine of the Trinity *ad intra* and *ad extra*, particularly in the act of creation, can the product of that act—the created world—be seen in its proper form. The universe can only be understood in trinitarian terms. For Christians, then, creation

---

40. Moltmann, *Trinity*, 102.

41. Ibid., 103.

42. Leo I, EP31, quoted in Edwards, *Jesus the Wisdom*, 55.

(the universe) has no independent history apart from the revelation of Father, Son, and Holy Spirit. (This is very different from the "common creation story," which can be discovered through any faith tradition as long as it is subordinated to contemporary science.)[43] More specifically, it is only through an understanding of God's activity through Christ in the Spirit that creation can be understood at all. That is to say, Trinity is our God, the source of all being, the foundation of all there is, and the Trinity is the mutually indwelling, self-giving communion of divine persons. The universe, *ta panta*, is thoroughly, essentially trinitarian.

This is true whichever model of the Trinity best expresses current understandings in the church. "Father, Son, Holy Spirit" emphasizes the salvation history of the Christian experience. It also emphasizes relationality: fatherhood in relation to Son, sonship in relation to Father, spirit "proceeding," or breathed from Father and Son. The identity of each person of the Trinity consists in their relation to the other persons; each is unintelligible except as connected with the others; all three are fundamentally connected and interrelated. Augustine's model of the Trinity as Lover, Beloved, and the Bond of love between them; Hildegard of Bingen's description of God as brightness, flashing forth, and fire; Bonaventure's terms Fountain Fulness, Word, and Love; the eternal Majesty of God, incarnate Word of God, and abiding Spirit of God; or Elizabeth Johnson's triple helix—all struggle to convey, with differing emphases, "the richness of holy triune mystery, inwardly related as a unity of equal movements, each of whom is distinct and all of whom together are one source of life."[44] The perichoretic dance of divine intimacy and mutual love overflows in creation and becomes expressed in the dance of the universe. The Trinity is characterized by supreme dynamic unity, supreme communicability, co-equality, and co-eternity. Therefore, relations of mutual love, persons

---

43. It is also different from "creationism," which takes Genesis 1 and 2 as a sort of freestanding explanation of the world, rather than as a narrative within trinitarian theology. As Reinhard Hütter writes, "'Creation' is only intelligible as 'doctrine' when it is part and parcel of the proclamation of the gospel, the redemptive story of God with Israel and in the life, death, and resurrection of Jesus of Nazareth," (Hütter, *"Creatio ex nihilo,"* 91).

44. The "eternal majesty" model is from *Enriching Our Worship*, 70. Johnson, *God For Us*, 221.

indwelling without loss of distinction, are the foundation of everything—not only human life, not only earthly life, but divine life as well, insofar as we comprehend God's self-revelation to us. Relationship is the primary metaphysical category. Elizabeth Johnson puts it this way: the God who is three times personal signifies that relatedness, rather than the solitary ego, is the heart of all reality. In this universal relatedness, mutuality and distinction are not opposites, but they require each other. Genuine mutuality of relationship involves dynamic distinctions. Contrary to what we frequently believe, difference does not imply either competition or inequality; instead, difference and love are equally characteristic of God and of the universe.[45]

CREATION EX NIHILO

The Genesis accounts do not explicitly assert that God created "out of nothing."[46] Some interpreters have refused the doctrine on this basis, viewing it as a late addition to Christian dogma. Peter Ochs, however, argues that the Hebrew of the beginning of Genesis is best translated, "In the beginning of [the activity of] God's creating heaven and earth [when] the earth was unformed and void . . . "[47] The question of any prior status of heaven and earth is left open; God may have already created the unformed "stuff" before shaping it into the universe. Creation *ex nihilo*, at any rate, is what Christians have attested, with few exceptions, at least since the time of Irenaeus. Jettisoning this would imperil Christian understanding of God's sovereignty and gracious providence, although it would probably not, in our scientific age, revive the particular pagan notion of the world as eternal.

Continuing to affirm creation *ex nihilo* is important to the relationship between creation and eschatology. The doctrine emphasizes the uniqueness of God's creating work: creating the universe is *not* the same as a sculptor shaping clay, or even a geneticist "producing" hybrid mice.

45. Johnson, *God for Us*, 216.

46. There are a few relevant biblical references: 2 Maccabees 7:28; Romans 4:17; Hebrews 11:3.

47. Ochs, "Genesis 1–2," 8. See also Eichrodt, "In the Beginning," 72–73.

Therefore, as we saw in considering God's creative freedom, the world we live in *and* the world to come each bear a strong element of mystery and surprise. Just as the natural processes and "laws" of this world originate in God's loving will, so does the nature of the eschaton.

Catherine Keller argues that creation *ex nihilo* means that God creates with an absolute, unlimited, dysrelational power, and that such a doctrine of God badly needs amendment.[48] But this is to imagine that there is a possibility of existence or power outside of God's agency, as if God somehow usurped what properly belongs to creatures. Rowan Williams explains this kind of fallacy:

> . . . creation is not an exercise of divine power, odd though that certainly sounds. Power is exercised *by* x *over* y; but creation is not power, because it is not exercised *on* anything . . . what creation emphatically isn't is any kind of imposition or manipulation: it is not God imposing on us divinely willed roles rather than the ones we "naturally" might have, or defining us out of our own systems into God's. Creation affirms that to be a part of this natural order and to be the sort of thing capable of being named—or of having a role—*is* "of God"; it *is* because God wants it so. And this implies that the Promethean myth of humanity struggling against God for its welfare and interests makes no sense: to be a creature cannot be to be a victim of an alien force (colonized by an alien "culture") . . . Creation in the classical sense does not therefore involve some uncritical idea of God's "monarchy." The absolute freedom ascribed to God in creation means that God *cannot* make a reality that then needs to be actively governed, subdued, bent to the divine purpose away from its natural course. If God creates freely, God does not need the power of a sovereign; what is, is from God.[49]

48. Catherine Keller, "Power Lines," 72.

49. Rowan Williams, *Christian Theology*, 68–69. One way to see this is to recognize the metaphorical quality of God's "sovereignty." The Bible's assertions that God is sovereign over the world do not mean that God is therefore limited in the way human sovereignty is. God's power is not taken or evacuated from any other source (such as humans, or creation); it was, is, and always will be singularly divine power. See also Hanby, *Augustine and Modernity*, 72.

The second point about creation *ex nihilo* is that because chaos is not something that pre-exists God, there is no force of evil that somehow escapes divine dominion. Rather, chunks of disorder temporarily mar the goodness of God's works. This is not to diminish the severity of evil, but to limit its scope: evil does not precede, overcome, or outlast the good.[50] All material existence has its ground in God: there is no other, competing, order of being.[51]

Third, assertion of creation *ex nihilo* reminds us that space and time also do not precede God, but are part of God's creation. Space and time impose finite limits on all creaturely existence, despite human attempts to transcend them. (Theology may be especially prone to these efforts, as it is all too easy to assume God's timelessness as our own!) Such attempts might well be viewed under the category of sinful pride—a rejection of the parameters God set on worldly existence. Because of our limited creaturely perception, we cannot see the world as it is for God—as it truly is. However space and time may figure in God's life, they clearly do not impose limitations on divine activity. Christofer Frey writes, "Everything possible relies on a reality which is hidden to our everyday perception."[52] The created world is but a limited, finite, fractional effect of God's trinitarian being. This means we should beware interpreting creation-redemption-salvation in sequential terms, so that eschatology is about the future while creation is about the past. Garrett Green argues,

> . . . eschatological visions in terms of an End *Time* cannot be taken literally, as though these events were scheduled to take place at some future point in time (whether or not that time can be known or predicted is irrelevant; the point is that the End, whenever it comes, is conceived in terms commensurable with past and present time). If eschatology were composed of prognostications about the future, such a compatibility of time and eternity would be appropriate. But since time and eternity are incommensurable, their relationship must be analogical rather than

50. Of course, this is part of Augustine's definition of evil as the privation of good. See also Haas, "Significance of Eschatology," 330.

51. Schwöbel, "God, Creation, and Community," 163.

52. Frey, "Eschatology and Ethics," 71.

literal; in other words, temporal imagery about "future events" should be read as metaphorical language about a discontinuous and inconceivable eternity.[53]

Green is right to caution against literalizing temporal language, especially in reference to God. Note, too, that when theology "temporalizes" God, it usually does so with post-Enlightenment categories of time and space, which are peculiar to recent Western culture. Western Christians do not control God's time, and neither do we control the world's time. Yet, Green's explanation overstates the case. Because God's action in creation takes place *within* the framework of time and space, it is important not to see the eschaton as timeless in the sense that it does not, and will not, have historical effects. God is not constrained by history, but God acts in history.[54] The creation of the universe, the incarnation of Jesus Christ, the resurrection—all happened in and to a temporal-spatial world. We have no reason to believe that the world's final consummation will be completely otherwise.

RESULTS OF GOD'S CREATING

❧ Thus far, I have shown that an understanding of creation as universes on their way to fulfillment in God agrees with traditional Christian understandings of God's creative activity. What, though, about the created world itself? Contemporary secular and "new age" views of the universe range from an evolving, self-guiding but ultimately self-destructing system (popular version of physical science) to a spiritual organism that yearns for mutuality (new age view) to a combined resource pool and laboratory for human endeavor (economic and industrial view). What can (mainline) Christianity offer as a corrective? In particular, how do Christian perspectives on creation match my description of creation as beings on pilgrimage? The next several sections argue that in its goodness, diversity, particularity, and in its corruption, creation exhibits its participation in God's intentions—that is, its eschatological character.

53. Green, "Imagining the Future," 81.
54. For a contrasting view, see Hardy, "Creation and Eschatology."

CREATION'S GOODNESS

To begin with, it is clear that the created world is good, although fallen and disordered. Genesis 1 declares seven times that God's creation is good. And because the speaker is God, earth's goodness is not simply described but *pre*scribed: "a divine word that must prevail."[55] On the other hand, as early as Genesis 6, "the earth was corrupt in God's sight, and the earth was filled with violence" (Gen 6:11). Yet, the corruption was insufficient to nullify all the earth's goodness. And so it has continued to this day. Earthly reality pulses in a constant tension between the goodness of its creation and the disorder of its Fall. It is appropriate, therefore, that one of Christianity's most consistent positions has been the qualified endorsement of the created world. Many recent books by Christian theologians affirm that Christians understand God values the created universe, and that, therefore, it is valuable.[56] What is more useful here is an analysis of how and why creation is valuable, how it is fallen, and what difference the Fall makes theologically.

CREATION'S DIVERSITY AND PARTICULARITY

Let us keep in mind that while we can make generalizations about God's creation, it is in fact comprised of a myriad of creations, each with its own character and particular destiny in God. It is often said that the idea of "nature" as a single category of creatures, elements, and forces governing their behavior was foreign to the Israelite's thinking. Land was the key concept, and they perceived everything else—climate, animals, plants, soil, and rocks—in relation to the land and its fertility. And God's will controlled the land, not a set of abstract mathematical laws.[57] So rather than seeing the universe as a single thing, Old Testament writers tended to see a myriad of creatures and forces united by their divine ori-

55. Clifford, "Bible and Environment," 6.

56. Many of these have been written to defend the tradition against either Lynn White's famous charge that Christianity is largely responsible for the environmental crisis, or the more generalized view that Christianity is a world-denying religion. White, "Historical Roots."

57. Brueggemann, *Land*.

gin and governance. What categories we see in the Old Testament match the perspectives of herding and farming cultures: garden is different from wilderness, and domestic animals from wild animals and predators.[58] Animals are rarely regarded simply as generic "animals," much less an undefined part of a generic "nature."[59] Similarly, animals and plants in Jesus' parables tend to behave according to their kind: "sparrows drop dead, dogs scavenge and lick the wounds of beggars, and eagles gather over a carcass . . . . There is a certain realism here, and a respect for the way things are, since that is the way God has made them."[60] And the way God made things is an amazing number of ways that humans have not even begun to understand.[61] This is a point at which recent secular environmentalism and Christian theology may concur: the caution against abstracting biophysical specificities into a single thing called "nature." As Friesen points out,

> we [humans] do not really experience abstractions like "biosphere" and "environment," especially when we are describing the experience of crisis. Rather, we experience concrete and particular changes within the horizons of a particular place—horizons which are never so broad as to be captured by the word "environment." So when we sound the alarm of crisis, what we mean to say is that something has happened to the places in which we live, that something has come between us and the fields and

---

58. The distinction between garden and wilderness carried profound implications for Christian treatment of their biophysical environment in the colonial era, when unfamiliar lands were viewed as ungodly wilderness awaiting transformation into blessed gardens. Not only was this approach disastrous for the land's prior occupants, it also distorted the Bible's own narrative. In both Old Testament and New Testament writings, both garden *and* wilderness were suffused with God's presence and under God's care. Biblical writers may have understood what subsequent generations forgot, that the garden-wilderness distinction holds with regard for human endeavors, but not necessarily for God's providential intentions.

59. Where the text reads "animals," the reference is usually to livestock, as the context makes clear.

60. Muddiman, "New Testament Doctrine," 30.

61. Job 38–39.

forests and the rivers, and between here and the mountains that
have marked for generations the places we call home.[62]

It is important for Christians to note that the diversity of creaturely experience (for many nonhumans indeed have their own perspectives) may
not match the diversity of God's creation. That is, we cannot presume to
know, to have mastered epistemologically, God's "map" of this created
universe. Racism is one of the more striking examples of attributing a human construct of difference to divine intent, with horrible consequences.
On the other hand, mistakenly identifying sameness can also have tragic
results. Gorillas, for instance, are much more similar to human children
than to geckos, yet they have been lumped together as "animals," and
seen as equally fit to live in cages under human scrutiny and for human
pleasure.[63] All of our characterizations of God's creations, no matter how
scientifically or theologically "certain," must be open to revision. What
we can say, based on trinitarian faith, is that God's creations differ and
converge in ways we do not comprehend; that differences do not necessitate conflict or inequality; and that all of God's creations are united in
their common origin and destiny in Christ.[64]

Another important point about creation's variety is that diversity
itself is good. Theologians have, from the early years of Christianity, said
that the diversity of creatures reflects the immeasurable richness of their
Creator. Creation is good because all created entities mirror, in some
way, the divine Trinity and its infinite goodness. The infinite distance,
however, between God's goodness and created goodness means that only

---

62. Friesen, "What are we Fighting For?" 48.

63. In addition to the many excellent studies of gorillas, a wonderful source of insight
on primates and other animals is Linden, *Parrot's Lament.*

64. Unfortunately, theologians continue to frame ecological issues as "humans" in
relation to "nonhumans" or even "nature" as if, even while acknowledging the difficulties
of the Cartesian dualism, they are simply unable to imagine beyond it. See for instance
Peter Scott and John Zizioulas. Interestingly, the tendency to think solely in human
versus non-human terms seem associated with less attention to the biblical texts. Both
Scott and Zizioulas are deeply concerned to explain the rationality of God's action (rather
than, perhaps, describe an appropriate response to it). Peter Scott, "Nature." Zizioulas,
"Preserving God's Creation."

a plethora of good creations can reflect, even imperfectly, the innumerable aspects of divine excellence. Typically, as in Bonaventure's work, creation's reflection is regarded as christological. The Word of God is the mediator or divine pattern of all things in the universe(s), so that creatures "seem to be 'nothing less than a kind of representation of the wisdom of God, and a kind of sculpture.' They are the work of art produced by divine Wisdom."[65] Similarly, C. S. Lewis describes Richard Hooker's vision of the world: "We meet at all levels the divine wisdom shining out through 'the beautiful variety of all things' in their manifold and yet harmonious dissimilitude."[66] Each creature, each of the innumerable bits of created matter, each community and ecosystem, retains and exhibits a minute portion of its Creator's bounty. In a famous passage at the end of *The City of God*, Augustine writes in awestruck praise,

> Shall I speak of the manifold and various loveliness of sky, and earth, and sea; of the plentiful supply and wonderful qualities of the light; of sun, moon, and stars; of the shade of trees; of the colors and perfume of flowers; of the multitude of birds, all differing in plumage and in song; of the variety of animals, of which the smallest in size are often the most wonderful—the works of ants and bees astonishing us more than the huge bodies of whales?[67]

It is important not to forget that creation is as much or more a creation and sustaining of relationships as it is of bodies. Here again, Christian theology and recent secular ecological work find agreement. Richard Hooker wrote, "God hath created nothing simply for it selfe: but ech thing in all thinges and of everie thing ech part in other hath such interest that in the whole world nothing is found whereunto anie thing created can saie, 'I need thee not.'"[68] The interdependence of creatures is not a new discovery; what is perhaps being *re*-discovered (in American culture, at least) is that humanity is not exempt from this interdependence. The great difference, of course, is that Christians believe these interrelation-

---

65. Bonaventure, *Hexameron*, 12, quoted in Denis Edwards, *Jesus the Wisdom*, 110.

66. Lewis, *English Literature*, 460.

67. Augustine *City of God*, XXII.24.

68. Hooker, "Sermon on Pride," 16–19.

ships to be not a self-generating blossoming of the earth, but as part of the gift of life from God. The "web of creation" is strung and restrung in the gracious care of the Holy Spirit. In Athanasius' words, ". . . it is [Christ] who has established the order of all things, reconciling opposites and from them forming a single harmony."[69] So as creation's beauty and diversity is christological, and Christ is the author and agent of ultimate peace, then creation reflects—in a dim and fractured fashion—its christic purpose.

THE FALL

Of course, creation is not pure goodness—neither the creatures nor their interrelationships are free from evil. If we define "Fall" in the narrow sense, the question is whether the sin of the first humans corrupts not only the rest of humankind, but the entire cosmos and its inhabitants irrevocably, until God's final saving action. Genesis 3:15–19 describe how Adam and Eve's disobedience affects nonhuman creation: enmity between the snake and the woman (and between their offspring), and a curse upon the ground—that its "natural" fruitfulness is diminished.

Most early Christian interpretations of the passage read it to mean that *all* of creation was perpetually damaged as a result of Adam's sin.[70] In particular, Theodoret of Cyr, Macarius, and Chrysostom, as well as the Cappadocians, took this line. For the Greek Fathers, the central idea of the human as microcosm entailed the cosmic effects of Adam's sin; because humans provided the "bond" for the universe, their actions shook the whole created order.[71] Irenaeus, however, understands the Fall specifically as willful disobedience against God—an act committed

---

69. Athanasius *Contra Gentes*, 113.

70. In the intertestamental *Life of Adam and Eve* (100 BC–AD 200), Seth is attacked by a "beast." When Eve chastises the beast, it retorts that its hostility is the result of Eve and Adam's own behavior. See Charlesworth, *Old Testament Pseudepigrapha* v.2, 273.

71. Louth, introduction to *Ancient Christian Interpretations*.

only by humans.[72] He believes that most non-human creatures have "persevered, and still do persevere, in subjection to Him who formed them." Overall, according to Irenaeus, non-human creation has retained its original goodness and order.[73] This argument occurred, of course, in the context of Irenaeus' refutation of Gnosticism, including the radical gnostic devaluation of the material world. It was critically important for Irenaeus' position that material creation was not considered second best to spiritual life.

Augustine, too, rejected the notion of a cosmic Fall. He wanted to assert the continued goodness of God's creation in contrast with the Manicheans. He also tends to read Genesis allegorically, so the thorns growing on earth represent the "prickings of torturous questions or thoughts concerned with providing for this life."[74] But creation itself was good, indeed full of delights and awesome pleasures, reflecting the glory of its Creator.[75] As Paul Santmire explains, Augustine believes the curse of Eden "touches human life alone; it does not disrupt the cosmos."[76] Augustine's rhapsodic praises of the natural world bespeak, perhaps, his own comfortable circumstances; at any rate, he seems far more sensitive to the beauty of creation than to its physical suffering.[77] Aquinas, following Augustine, also rejected any corruption of the non-human world. "For

---

72. I beg the question of whether or not any animals commit sin, which I view as a serious question. In a wonderful passage about dolphins (the mammals, not the fish), Michael Northcott writes, "Do dolphins sin? That is, consistently fail to realize their flourishing, fight wars, become addicted to destructive behavior patterns, etc.? Apparently they do this much less than humans and some other animals. They are of course affected by human sin, especially by industrial fishing practices . . . . Dolphins don't carry the cross, as far as we know, because they haven't heard the Gospel. But they do share in the hidden meaning of reality which the cross shows forth, and they're exemplars of the moral priority of the weak . . . part of the praise of the creator," ("Do Dolphins Carry the Cross," 13). See also Bauckham, *God and the Crisis*, chapter 7. One thing is certain: animals are part of creation's witness and worship of God, simply by being themselves.

73. Irenaeus *Against Heresies* 2.28.7.

74. Louth, introduction to *Ancient Christian Interpretations*.

75. Augustine *City of God* XII.4.

76. Santmire, *Travail*, 66.

77. See Augustine *The City of God* XXII.24 for a good example.

the nature of animals," he says, "was not changed by man's sin, as if those whose nature it is now is to devour the flesh of others, would have lived on herbs, as the lion and the falcon."[78] In subsequent centuries, Christian attitudes toward the post-lapsarian status of creation varied, although the view that creation had escaped the Fall was typically a minority opinion, especially after the Reformation. Either view raises serious difficulties. On the one hand, to suppose that human beings managed to corrupt the whole creation that was otherwise in good order, conforming to God's intentions, seems incredibly anthropocentric. It also knots up any attempt to view God's creative activity as a gradual, developmental process. Our choices seem to be: either millions of years of life got derailed by those disobedient humans, or else the Edenic narrative must be completely ahistoricized. On the other hand, to suppose that only humans suffer the consequences of human sin is manifestly false, especially from an ecological perspective. Such a view also introduces its own sort of alienation between humans and nonhumans, as if the moral status of humans is disconnected from their bodily existence.

It is not necessary, however, to resolve this question here. The significant matter is not whether earth is corrupted by human sin, but whether or not earth is corrupted at all. That is, we can broaden our definition of "fallen" and ask whether creation is other than it should be, disordered, less than ideal, or contrary to God's purpose and plan. "Fallen" in this sense is different than finite, at least logically different. Most Christian theologians hold that the finitude of the world is part of God's creative intent. The understanding that the universe and its creatures have a beginning and end was an important component of Hebrew and Christian faith from the earliest times. That God created this world and is in charge of the world's end sharply refuted both ancient ideas of the cosmos being in some sense divine, and the Greek idea of earth's eternity. Anything created is not God and therefore not perfect or eternal. It is possible, however, to imagine a finite world in which lives are limited—a world in which creatures are born, they live peacefully, and die—but without the vast conflicts and suffering that life entails. The question of earth's im-

78. Aquinas *Summa Theologica* I.96.1.

perfect status is, then, twofold: First, are human observations of earthly disorder shaped by anthropocentric sentimentality? Maybe the creatures are not really suffering in the way we imagine, or the conflicts are not harmful. Second, even if suffering and conflicts are real, are they necessarily contrary to God's intention? Perhaps they are a corollary to finitude; everything takes up space, after all, and every living thing has to eat. If scarcity is endemic to finite life, as the argument states, then the same is true for conflict and competition.[79]

To answer the first question (whether nonhumans really suffer), the *prima facie* evidence for non-human suffering (otherkind) is readily available: creatures die by freezing or starving; animals eat each other (often alive); hurricanes, tornadoes, fire, and droughts wipe out human and non-human communities. For most of human history, people have assumed that animals experience the sensations of pain, pleasure, and even emotion that they exhibit behaviorally. The Old Testament, in particular, prohibits certain treatment of domestic animals *because* it would hurt them. The Sabbath rest, for instance, applies to donkeys as well as humans, and fallen livestock should be helped to their feet (Deut 5:14, 22:6; Exod 23:5). In fact, the Old Testament implies the possibility of animal psychological suffering, as well: the only reason an ox should not be muzzled while threshing grain is that he might become frustrated by not being able to eat what is right under his nose (Deut 25:4).[80] New Testament texts rarely treat non-human nature explicitly, but certainly do not contradict Old Testament instructions regarding animal welfare. Jesus said, echoing earlier testimony to God's universal providence, that God feeds the birds of the air (Matt 6:26) and attends to individual sparrows (Luke 12:6). The biblical witnesses, therefore, assume the possibility, existence, and moreover, the undesirability of animal suffering.

This belief persisted through Western culture's medieval era. It began dwindling in the early Renaissance, and became a minority viewpoint in

---

79. Richard L. Fern argues somewhat along these lines in *Nature, God, and Humanity.*

80. The questions of meat-eating and animal sacrifice are too large to be addressed in this work. The point here is that animals under human possession should actually be cared for, not simply owned and worked.

the seventeenth century.[81] Anthony Le Grand's assertion that dogs under torture feel nothing, but scream merely as an insensate reaction, would have sounded as bizarre to his predecessors as it does to most of us.[82] Yet, it was a necessary move in a general humanistic shift. For humans to regain their "dominion" over non-human nature—as Bacon wanted and thought was divinely intended—"man" needed to subject nature to probing tests that would wrest "her" secrets from her. A good God could not allow such suffering as this would require; therefore, the appearance of suffering on the animals' part must be an illusion. Aside from the grotesquerie of this logic, it is interesting that even here, the Fall of creation was not denied. Rather, Bacon and his peers interpreted it to mean that humans were improperly subject to the destructive forces of nature.

In sum, the arguments on the side of creation's corruption are persuasive, whatever the specifics of how that corruption came about.[83] Moreover, it is clear that such corruption contradicts God's intentions for the world. Recall, first, the Old Testament admonishments to treat domestic animals kindly. Then consider the repeated assertions by the Old Testament prophets that the Israelites' unfaithful behavior bore severe consequences for the land.[84] If drought or famine is a penalty for sin (even a God-given penalty), it cannot be desirable in itself. Creation's suffering and disorder are indications of matters gone wrong in the created world. Finally, as chapter 4 argues in depth, God's promises consistently include not only justice among people, but abundant life for all creatures and harmonious relations between species. These promises reverse the present order *because* the present order contradicts God's ultimate intentions.[85]

81. Thomas, *Man and Natural World*.

82. Le Grand was one of Descartes' followers, many of whom shared this view, although Descartes himself was less radical. See Le Grand, *Entire Body*, ii.252, cited in Thomas, *Man and Natural World*, 315.

83. In the early church, much of what we would call natural evil as well as injustice was attributed to demonic powers that gained a foothold on earth through the Fall of the angels. See Galloway, *Cosmic Christ*, 28.

84. See, for instance, Amos 6:4–14 and Northcott, *Environment and Christian Ethics*, 187ff.

85. For an opposing view, see Fern, *Nature*.

In Allan Galloway's words, "the symbol of cosmic fall is the complement of the symbol of cosmic redemption."[86]

To put it another way, creation did not avoid the Fall; non-human creation suffers corruption and distortion just as humans do, even though both remain within the compass of God's love and care and are, therefore, good.[87] Why, though, does it matter whether or not creation is disordered? What is at stake, especially in terms of eschatology? It turns out that a lot is at stake for Christians, and this is another point at which we must depart from the "common creation story." Let us look first at the alternative: suppose this is the "best possible world," or at least a world that is not deeply corrupted. Then what we think of as "natural evil" must not only be fundamentally different from human sinfulness or social evil, but it also must not be evil at all. "Natural evil," instead, is simply part of God's intentions for the world. This goes even farther than the Calvinist understanding of natural evil as God's judgment against sin; as we noted above, the existence of suffering and death as punishment does not entail or even imply the independent value of suffering and death. The idea that suffering *in and of itself* is good "resolves" questions of theodicy by dismissing them altogether, in defiance of vast experience and biblical testimony.

On the other hand, understanding non-human nature as corrupted, though good, makes important theological points. It underscores that "the very nature of creation is always ambiguous; it points both ways; it affirms and denies God at one and the same time. It affirms God because God loves and cares for it but it also necessarily denies God because it is not divine."[88] This pervasive ambiguity of nature also means that natural and human evil are never completely distinguishable. What humans do matters—has meaning—in ways that transcend the personal, social, political, or "spiritual" realms. It matters in and to the world beyond human culture. Conversely, humans are *necessarily* subject to the conditions of earthly life. Whether her sins are horrible or minor, any human being will suffer the effects of natural corruption and death. This is both com-

---

86. Galloway, *Cosmic Christ*, 21.

87. Rom 8:19–25.

88. Linzey, *Animal Theology*, 81.

forting and sobering news. It is sobering because nothing in the created world, no environmental ethic, no human power, no evolutionary shift, can "cure" creation. "A new and special act of God, accomplished in the death and resurrection of Christ, is necessary to purge creation of the corrupting effects of sin."[89] In this rather backhanded way, creation is shown again to be eschatological; none of God's creatures can in themselves fulfill God's purposes for them. Yet, it is comforting to realize, as Stephen Webb says, "that nature is fallen, and that nature's distortions transcend the many harms humans contribute to it, so that we should deplore the human abuse of nature but we should not think that humans cause all of nature's problems."[90] This is a subtle point. Overemphasizing the corruption of creation (as some may claim of Calvin and, in his wake, Jonathan Edwards) can lead toward an undervaluing of the biophysical world, and thus an abdication of human responsibility for the effects of human behavior on the rest of creation.[91] We see this attitude in some "wise use" literature: if species extinctions happen anyway, why are human-caused extinctions any cause for concern or remorse?[92] Obviously, this argument can neither withstand logical scrutiny (the existence of evil does not generally justify the commission of evil acts), nor, from a Christian perspective, adequately describe God's relationship with all of creation. One of the fundamental tenets of Christian faith, after all, is that God loves *despite* the imperfections of the beloved. Rather than the beloved's corruption impeding God's charity, his charity is precisely directed toward the beloved's sinfulness and corruption so that grace can sanctify and redeem the beloved. On the other hand, underemphasizing the fallen character of creation is also dangerous. It can yield a romantic view of "nature" that contrasts

---

89 Haas, "Significance of Eschatology," 339.

90. Webb, "Ecology vs. Peaceable Kingdom," 240.

91. Tied into his stress on the corrupt character of creation is Calvin's implicit view that creation serves more to demonstrate God's justice than God's charity.

92. The "wise use" movement is a quasi-Christian anti-environmentalist movement that prioritizes financial profit and individual political freedom over state regulation or restraint. It characterizes all environmentalists as pagan, extremist, and seeking excessive State control of citizens. Ron Arnold coined the term in the late 1980s. See Burke, "Wise Use Movement."

the goodness of the nonhuman with the sinfulness of the human—at once reinscribing the human-nonhuman dualism and misrepresenting the complex character of the non-human world. This romanticism parades itself as a nature-friendly, sympathetic perspective, yet it diminishes the very real struggle, suffering, and waste that constantly occur in nonhuman creatures. In a sense, it denies God's care of the universe because it denies the universe's need for that care. Both errors, then, produce the same result. Either over-emphasizing or under-emphasizing creation's Fall refuses the expansiveness and depth of God's unlimited love for the world, and therefore nullifies the eschatological dimension of creation. Creation is neither ideal nor completely corrupt. Montefiore writes, "Christianity doesn't teach that this is the best of all possible worlds. It teaches rather that God's potentially good creation needs to be redeemed and sanctified, and that these are costly processes."[93]

In sum, the fundamental Christian characterization of God's creation is that it reflects and partakes of the goodness of its trinitarian Creator, but the reflection is currently dimmed and distorted, and can never be restored by natural means. As a result, the extent to which the current natural order can function as normative is limited. This is the difficulty with some natural-law approaches to ecological theology, such as Michael Northcott's *The Environment and Christian Ethics*. Northcott is quite right that "the evacuation of purposiveness and moral order from nature independent from human willing and purposiveness" from the time of Descartes and Bacon generates (or at least strengthens) the view that humans can dispose of the natural world in any way they like.[94] Any attempt, however, to infer moral order from empirical observation of human and/or non-human life faces the literally superhuman task of distinguishing the fallen, corrupt aspects of creation from the good, grace-filled ones, and fixing those distinctions in a way that denies both creation's dynamism and God's continuous participation in earthly life.[95] To make such distinctions, we must fall back on particular narrative

93. Montefiore, *Man and Nature*, 36.

94. Northcott, *Environment*, 68.

95. For an objection similar to mine, see Linzey, *Animal Theology*, 81.

traditions—which immediately contradicts the alleged universality of empirical observations. Gender roles provide a classic example. Does the common—perhaps ubiquitous—subordination of females in human societies point to an order intended by God or to a common effect of human sin? Or some combination of both? We can only answer these questions from a position inside the Gospel witness, where the "universality" of gender roles becomes relativized or immaterial. Additionally, while the Western Enlightenment clearly played a major role in producing the current environmental crisis, harmful behavior toward non-human communities has been just about as common as gender subordination, and likewise, as resistant to resolution by natural-law approaches.

CREATION AS ESCHATOLOGICAL

Northcott, though, rightly points to the centrality of purposiveness in any ecological theology. God's creation is suffused with purpose and promise: all of creation, including humans, is made for the glory of God. In the words of Steven Bouma-Prediger, "God is the center of the cosmos, and our [human] task and privilege is to worship the Maker of heaven and earth in concert with all other creatures."[96] I take it as given, here, that non-human creation does not exist solely to assist in human salvation, or to chastise "man," as has been argued occasionally in the Christian tradition. Biblical testimony as well as ancient and contemporary theologians argue decisively that non-human creation serves God, praises God, and will partake in the Kingdom of God.[97] All creatures, living or nonliving,

96 Bouma-Prediger, *For the Beauty of the Earth*, 174. Saint Basil prayed, "O God, enlarge within us the sense of fellowship with all living things, our brothers the animals, to whom you gave the earth as their home in common with us. We remember with shame that in the past we have exercised the high dominion of man with ruthless cruelty, so that the voice of the earth, which should have gone up to you in song, has been a groan of travail. May we realize that they live not for us alone but for themselves and for you and that they love the sweetness of life." Quoted in Wynne-Tyson, *Extended Circle*, 9.

97. The point that God's care for non-human creatures transcends humanity was one of the first arguments by ecological theologians in this century. The argument rests soundly on biblical and early church sources, which have been largely ignored in the centuries since Descartes.

sentient or not, live in doxological response to God's gift of creation and promise of redemption.[98]

> Let the heavens be glad, and let the earth rejoice; let the sea roar, and all that fills it; let the earth exult, and everything in it. Then shall all the trees of the forest sing for joy before the Lord; for he is coming, for he is coming to judge the earth. He will judge the world with righteousness, and the peoples with his truth. (Ps 96:11–13)

The book of Revelation also paints wonderful images of "every creature in heaven and on earth and under the earth and in the sea" singing praises and worshipping the Lamb at (or even "after") the end time (Rev 5:13–14). Chapter 4 will address these images of the eschaton in more detail; the point here is simply—though crucially—that the purpose of creation is the glorification of God, and that purpose is fulfilled partially in earthly life, and only completely in the final consummation. Thus, the universe is doxological and eschatological in its very being.

The details of this claim are important. Christian eschatology differs from Aristotelian teleology in that its telos is granted as a gift from God. Scientists speak of the "inner" drive of an organism to survive and reproduce, but this is different from creation's eschatological goal, which transcends the organism, species, and even the category of earthly life. As Conyers explains, "the world does not find its purpose in itself. It is "eccentric" because it centers outside of itself; it centers in God . . . ."[99] This is another point where the "common creation story" diverges from a Christian perspective.[100] The universe, then, is not a closed system; it is always open to the gracious presence of the Spirit in accord with

98. Students often react scornfully to the story of Saint Francis preaching the Gospel to birds. However, if the hills and valleys can sing God's praises, and if we understand "preaching" as more than making oral propositions about God, it seems that preaching to nonhumans may well be a "prudent" Christian act. The harder task for most humans (except Francis) is to listen when the hills and valleys and birds preach to *us*.

99. Conyers, "Living Under Vacant Skies," 15.

100. Wilkinson describes ways people have tried to reconstruct a wholly immanent telos to the universe in "New Story of Creation," 26–36.

the trinitarian God's intentions. What I describe here is very close to the traditional idea of divine providence; yet, I want to emphasize that *the entire* universe, not just human lives and interests, is under God's providential care.

The strong connection proposed between material creation and its creator may seem worrisome. This chapter insists that God's creative activity and sustaining presence are not intermittent "descents" into creation, but they are continuous and pervasive participation by Godself; therefore, all creation, by virtue of divine grace, partakes of divine goodness. How, then, is the universe not an extension of Godself but rather a creation? The universe is creation, not emanation, in part because of God's absolute freedom in creating, and in part because "the action of God as creator constitutes things as beings in their own right."[101] God creates through love that which is not God: a contingent, spatiotemporal universe of beings in relation to each other and to Godself. Moreover, God has granted the universe a *kind* of independence through its "natural" processes and events. Richard Hooker's understanding of law may be helpful here. For Hooker, the regularity of natural phenomena—the laws of nature—reflect both God's faithfulness in sustaining life, and God's graciousness in granting every thing a "natural course," a way of being that is particular and beneficial to itself.[102] Otherwise, the universe would be unintelligible.[103] In this way God exercises both boundless love and divine restraint.

Furthermore, the theocentricity of the biophysical universe reaffirms, rather than undermines, its distinctiveness from God. God creates, sustains, and redeems that which is distinct from the divine, ultimately to draw the universe back into his own life. George Hendry explains this as follows:

101. Gunton, *Christ and Creation*, 90.

102. Hooker, *Lawes of Ecclesiastical Polity*, Book I.ii.iv, 203ff.

103. John Polkinghorne is a good source on the related issue of the dependence of science on theology; his work is also useful for reframing the "laws" of nature in terms of recent quantum physics. See Polkinghorne, *God of Hope*.

The mystery of nature follows the same dual movement of the Logos, since it is by the same Logos that all things were made . . . it is the mystery of Spirit that it loses itself in its opposite, and fulfills itself by bringing its opposite to fulfillment in itself . . . The world of nature is a world of matter and energy, force and gravity, space and time, contingency and irrationality, frustration and futility. Why did God create the world so different from all that he is believed to be? But the character of spirit is to go out from itself, to embody itself, to lose itself in that which is remotest from itself, in order to unite that other with itself.[104]

The relative independence of the universe from God allows the possibility of evil, suffering, and denial of God's grace—this is all true. Yet, it is *only* because creation is distinct from God that it can receive the gift of God's presence and final communion.[105]

SUMMARY

❧ This chapter argues that a "common creation story" differs substantially from the Christian creation story, and that the latter portrays a much richer and more complex image of God's relationship to the created universe. God's creating activity is absolutely free, gracious, trinitarian in character, continuous, and out of nothing. Creation itself is good because it reflects the goodness of its creator, yet it is also fallen and subject to evil and suffering. Creation is incredibly various and particular, with all things in the universe held in relation to each other and to God through Christ. Finally, it is eschatological: the Spirit sustains it until the fulfillment of all things comes with the Kingdom of God. This last claim cannot be overstated: it is not as though a self-sufficient entity (creation) pre-exists that plays a role in God's redemptive purposes. Rather, it is the case that God's providential care and intentions call the creation into being and sustain it until it is drawn fully into its consummated relationship with the Trinity.

104. Hendry, *Theology of Nature*, 171.
105. This is another problem with the emanating model of creation.

The next chapter will explore what difference these claims make for ecological theology and Christian practice: why it matters that the land we save (or do not save) is part of God's plan for the eschaton

CHAPTER 3

ETHICS AND ESCHATOLOGY

THE TASK of this chapter is to display a conception of the Christian life as witness: following Christ in proclaiming and demonstrating the Kingdom of God. The idea of Christian ethics as discipleship is not original to me, of course; it stems from the Gospels' record of Jesus' commands and from Pauline epistles. As a model for Christian life, individuals and communities throughout the past two millennia have exemplified witness, even though it has not been the dominant model in the last several centuries.[1] By witness I mean a particular understanding of discipleship in which the communal lives of the disciples testify, through character, worship, and action, to the Kingdom of God as inaugurated, preached, demonstrated, and promised by Jesus Christ. The goal here is not to prove that witness is the best single perspective on Christian ethics. Such an effort would risk not only obscuring useful alternatives, but also elevating an ethical perspective over the Triune God. Adopting "witness," however, as the primary ethical perspective performs several key tasks: it allows for a way out of the dilemma between messianism and despair in our ecological lives; it ensures a characteristically Christian unity of faith and action; it demonstrates the lived-out connection between creation and redemption; and it facilitates a sensitivity to differences in social location that shape particular Christian lives. These are crucial advantages for Christian ethics, especially in the context of environmental destruction.

The reader should be warned that this sketch of Christian witness may seem rather general; except for a few examples, I do not say what Christians ought to do in specific circumstances—namely, the current

---

1. Notable examples include St. Francis, the Anabaptists, and Dorothy Day.

circumstances of massive environmental degradation. The next chapter (chapter 4) addresses a number of specific forms of creation-focused witness. In fact, it is rarely possible to dictate "universal" practices for Christian witness apart from the specifics already mandated by Jesus himself: feed the hungry, console the afflicted, love one's enemies, preach the Kingdom. Christian witness is always a response to what God is already doing in particular situations/communities/struggles, so that, outside of those particular situations, witness can only be described in general terms. On the other hand, Jesus' life itself presents the epitome of "Christian living." As the Gospels teach, "loving" means serving the needs of others—even at the cost of humiliation, rejection, or death. The inclination to view Jesus' commands as too vague to follow, therefore, arises partly from the pietistic emphasis on individual emotive connection to God, and partly from selfishness. It is easier to identify oneself as a generally "loving" person than to undertake the uncomfortable, self-sacrificing work of feeding, housing, and serving particular poor people, prisoners, and enemies. Before further consideration of the benefits of witness, however, this chapter examines the object and character of a witness-approach to Christian life.

The connection between "witness" and the object of that witness—the Kingdom of God under the reign of Jesus Christ—is internal. That is, Christians witness to the Kingdom because Jesus *as Lord* witnessed to the Kingdom. The Kingdom is not a generic ideal that Jesus happened to talk about during his ministry, but the realization of his redemption of the world. And redemption is another way of describing "bringing back to God." So Christians witness to Christ and his work of ultimately returning all of creation back to God; that return, or communion, is the Kingdom. I begin by addressing epistemological questions: given the very nature of the eschaton as being beyond anything in earthly experience, what can Christians know? It turns out that while our knowledge of the Kingdom stems from a traditional source—Scripture—it requires a very particular mode of learning. Knowing God's Kingdom, that is, is a way of knowing God, and knowing God has always required particular practices or disciplines, self-commitment, and faith. Moreover, because the Kingdom of God is the *new* creation that redeems the *old* creation, it is important to comprehend how memory shapes Christian perspectives of the eschaton. Ironically, perhaps, we cannot grasp the reality of the eschaton without a comprehensive theological grasp of the present and past's reality, including the present and past of the non-human world.

I explore the character of the Kingdom itself—what the Old Testament prophets declared, what Jesus preached and demonstrated in his ministry, and what emerges from Christian understandings of the Triune God. This material will likely be familiar to readers, except perhaps in its implications for non-human creation. After all, peace, abundance, and justice are promised to more than just human communities. Finally, I argue for witness as a normative description of Christian life. This description is determined by the nature of the Kingdom itself and by the life, death, and resurrection of Jesus Christ. Therefore, witness is characterized by obedience, self-sacrifice, hope, joy, and a faithful confidence in God's tendency to explode the boundaries of the possible.

This chapter draws on a variety of sources, some of whom would not ordinarily find themselves in conversation, for instance, Richard Hooker, Jon Sobrino, and Christof Schwöbel. Yet, in ways I believe to be true to their individual convictions, they all contribute to the argument that living in witness to the eschaton is the way Christians follow Jesus Christ.

"KNOWING" THE KINGDOM

❧ Witnessing to the Kingdom of God is not as simple as it might appear. Christians must understand what we are witnessing to—what do we believe about the Kingdom? What grounds these beliefs? How does (and should) social location bear on our beliefs? After considering these questions, it will be useful to explore the variety of language used to express the relationship between human lives and the Kingdom of God. Why and where is "witness" a better term than "building" the Kingdom? What is at stake in the language Christians use? Then it is important to unpack the understanding of witness itself. What are the key characteristics of witnessing? How can our stumbling through the ambiguities of earthly life possibly point to the reality of God's promises?

KNOWING THROUGH PRAXIS

The first point, of course, is that we do not *know* anything about the Kingdom of God in the epistemological sense of scientific observation or "objective" experience. But it is crucial not to misunderstand this caveat. From a theological perspective, even if Western science could elucidate aspects of the Kingdom, it would not "prove" its reality. The lack of laboratory evidence, conversely, should not undermine our faith in the reign of God. It is better to regard our ignorance and uncertainty about the Kingdom as resulting not from the ambiguity of the promise, but rather from the distance between Creator and creation, from the transcendence of God beyond all earthly understanding, and from the limitations of human sin. As Richard Hooker wrote,

> Dangerous it were for the feeble brain of man to wade far into the doings of the Most High; whom although to know be life, and joy to make mention of his name; yet our soundest knowledge is to know that we know him not as indeed he is, neither can know him: and our safest eloquence concerning him is our silence, when we confess without confession that his glory is inexplicable, his greatness above our capacity and reach. He is above, and we upon earth; therefore it behooves our words to be wary and few.[2]

On the other hand, Hooker himself is an excellent example of the counsel that speaking (or writing) cautiously does not mean refraining from speaking at all. Christians are commanded, after all, to spread the gospel, to speak the good news. Spreading the word about the Kingdom of God relies not on the confidence in any human capacity for perfect knowledge, but on the confidence in the perfect faithfulness of the divine object of our speech.[3] So to refuse all "speculation" about the Kingdom on the basis that biblical visions are too far removed from us, too metaphorical, or too incredible, is not an adequate response. Christians should be keenly aware of the mismatch between their human convictions about

---

2. Hooker, *Lawes of Ecclesiastical Politie, Polity* I.ii.3 citing Ecclesiastes 5 (London: MacMillan, 1865) 201.

3. Rowan Williams, "Hooker," 373.

the Kingdom and its true glory; nonetheless, they must testify to these convictions as part of their faith. Garrett Green explains:

> The language of the Nicene Creed is significant; "and we *look* for the resurrection of the dead; and the life of the world to come." This is not the language of theory but of expectation, not of prognostication but of prophecy. The point is not that as Christians we have some special source of information unavailable to people generally, on the basis of which we claim to know beforehand (prognosis) what is going to happen in the future. Rather, we confess our faithfulness to the vision of the world to come contained in, and implied by, the witness of the prophets and apostles in Scripture. Our confidence is grounded, *not in superior knowledge or insight, but in trust of those* whose imaginations illumined and captured our own.[4]

The second epistemological point is that understanding of the Kingdom is unattainable as an impartial, purely intellectual exercise. Certainly we study the biblical texts, and the church's reading of those texts, but as Jon Sobrino writes,

> What must be analyzed is the hermeneutic value of praxis—praxis as a means of grasping the nature of the Reign of God, in such wise that, conversely, without praxis an understanding of the Reign of God would be crippled and diminished . . . . Taking up and adopting a reality enhances one's grasp of the reality to be taken up and adopted.[5]

This is a very old Christian idea—that faithfulness to God is inextricably tied to knowledge of God; in fact, faithfulness is often prior to knowledge for God's creatures. It is only in following Jesus that Christians begin to understand what it means to follow Jesus. It is only in praxis (to use Sobrino's term) that Christians get a glimpse of the possibilities of the Kingdom. It is only in the praxis of eco-discipleship, therefore, that Christians begin to grasp the reality of God's redemption of *all* creation.

4. Green, "Imagining the Future," 78–79.
5. Sobrino, "Reign of God," 64.

This chapter offers, therefore, merely a starting point—a beginning analysis of what Christian ecological witness would entail. Such an analysis will deepen and change as it is taken up and testified to in practice.

Sobrino argues that this happens both positively and negatively. The praxis of serving the poor, of bringing hope to the downtrodden, reveals the object of that hope as the Reign of God, where love and justice are perfectly fulfilled. On the other hand, the praxis of resisting oppression brings one face to face with the anti-Reign, the existing structures of oppression and injustice. Sobrino writes, "it is in praxis, and not in the pure concept, that the existence and reality of the anti-reign appears with greater radicality." [6] Moreover, it is only in praxis that the cross begins to loom as the nearly inevitable outcome of Jesus' path, and therefore of the disciples' paths. Following Jesus brings one into confrontation with the anti-Reign, and triggers its defensive mechanisms. In Sobrino's words, "the anti-Reign inevitably produces persecution and death, because the God of life—the God of Jesus—and the gods of death are locked in mortal combat."[7] A better way of phrasing this, perhaps, is that the forces of the anti-Reign defend themselves with violence and death because those are the only means they understand. This is why the cross is central for eschatology as well as for ethics: in Jesus Christ's death and resurrection God received the strongest blow the anti-Reign could deliver and then rendered it ineffective. Nonetheless, the Kingdom has only been inaugurated, not fulfilled, so that disciples continue to run the risk of being taunted, threatened, persecuted, or killed. Only when God establishes the Kingdom, when the Reign is fulfilled, will death be vanquished entirely.

KNOWING THROUGH MEMORY

Our understanding of God's Kingdom is shaped not only by praxis, but also by memory. So an eschatological discipleship, such as I advocate here, seems somehow connected to personal and communal memory. Yet, why should ecology or eschatology be concerned with memory? Ecology, one would think, is acutely about the present—the temporal

6. Ibid.
7. Sobrino, "Spirituality," 249.

breakdown of biophysical systems in a finite world. And eschatology, of course, is about the future (or, to avoid over-historicizing, about the not-here-and-now), precisely about the future not being limited by past or present. So what difference does memory make to eschatology? What difference to ecology? The second question is more easily answered. Much ecological thinking, whether or not it is Christian, mourns an earlier time before the recent devastations humans commit upon non-human nature. Sometimes the "golden age" (or "green age") is seen as a generation back, when songbirds, frogs, and particular butterflies were more abundant in a favorite landscape. (Indeed, many environmentalists have this sort of experience; the recognition of particular, place-ful loss is often part of the development of contemporary pro-environmental attitudes, at least among those privileged enough to have had relatively nontoxic childhoods.) Sometimes the green age is anything prior to the Industrial Revolution; sometimes it is pushed back to hunter/gatherer societies; sometimes it is Eden itself.[8] The goal, then, of environmental efforts is shaped by the discursive memory of the "green age." Can frog populations recover? Can butterfly habitats be restored? Can the vistas of the Blue Ridge (or Denver or Mexico City or wherever) be cleansed of smog to regain their former clarity?

Our understanding of the enormity of recent ecological harm forms a necessary element of imagining and praying and working for alternative scenarios. We should, however, be alert to the gaps and distortions of our green-age images. Environmental destruction did not begin with the industrial era, much less in the last generation. King Solomon imported huge numbers of cedar logs from Lebanon; Imperial Rome was notorious for ravaging forests; China suffered horrible floods from deforestation in the tenth through twelfth centuries. Christian environmentalists tend to forget these facts, just as we forget that non-human nature—at least, post-Edenic nature—is disordered; it is not a peaceful, harmonious world

---

8. Ponting, *Green History*. Ponting identifies the shift from sustainable to unsustainable human living with the beginning of settled agricultural production.

even apart from human interference.[9] Now one could argue that, while our nostalgia for a "green age" may be inaccurate, such glossing over past reality is scarcely harmful. In fact, the image of our past (or nature's past) needs to appear highly desirable precisely because only its desirability can combat the fatigue and despair generated by our images of the present. Moreover, this book argues for the importance of Christian eschatology, or a vision of the future that can sustain our current discipleship in a time of little environmental "progress."

Nonetheless, we cannot forget for a moment that eschatology is not merely a religious visioning exercise, in which we conjure up inspirational imagery. While the biblical witness to the eschaton is multivocal, it is not a free-for-all; not anything and everything qualifies as heaven, regardless of its motivational power. Just as any other area of constructive theology, eschatology must be measured over and over against our best understandings of God's purposes as well as against its practical effects among believers. We would reject, for example, a vision of heaven in which almsgiving was rewarded regardless of the intention of the donor, even though that vision might result in more charitable donations by churchgoers. The present, practical benefit—more donations to the poor—does not outweigh the fact that, according to the prophets, the parables of Jesus, and Christian traditions, giving with an eye to reward does not correspond to the Kingdom of Heaven. Therefore, Christian environmentalists are cautioned against too easy a fulfillment of our desires for the non-human world in the eschaton. The Kingdom of God is God's purpose and activity, not our own.

Moreover, because nature is disordered, the eschatological reserve applies not only to human history, but also to the non-human world. The failure of anything this side of the Kingdom to match the glories of the Kingdom critiques ecosystems and interspecies relationships as well as political and economic systems. All creation groans for redemption not

---

9. The term "disordered" indicates that nature bears the effects of sin and finitude however one narrates the specifics of the Fall (see chapter 2). Eve Vitaglione, biologist and long-time leader of nature workshops for children, says that "everything is something else's lunch." This is true, but Christians should not regard it as an unqualified good (though it is better, thank the Creator, than everything being something else's poison).

only because of human sin, but also because of "natural," or non-human, evil. No scenario in nature, no matter how unpolluted by human activity and no matter how healthy in ecological terms, is the Kingdom of God. This second point is different from the nature-worship so decried by some Christian theologians. Here the error is not regarding nature as divine, but as not requiring any improvement. It comes from elevating the status of "natural" to normative, so that whatever happens apart from human intervention is the best possible world. Yet, inter-creature relationships produce real suffering, tremendous waste, and horrible loss of life.

From the standpoint of contemporary environmentalism, this rejection of romantic nature has two crucial implications for ecotheology. First, there is no "green age" for Christians to remember and restore. Hunter-gatherer societies laid waste to less acreage, certainly, but did not preclude death and misery. The ecological past, like the socio-political past, bears wounds, open sores, and dung heaps, as well as frolicking otters and beautiful vistas. Our hope must spring not from romantic wilderness landscapes, but from the resurrection of Jesus Christ in whom all things are made new. Second, in a Christian view, humans are not the only problem plaguing the universe. While the deadly environmental effects from human greed, sloth, impatience, fear, and so forth are undeniable, natural evil exists apart from these. Were the human species to disappear suddenly or—even more miraculously—to stop all sinning, suffering would continue on earth (although the patterns of suffering would certainly shift).

Here one might argue that in critiquing the role of memory in environmentalism, I seem determined to undermine the whole environmental project. If the world is a dreadful place whether or not people live in less wasteful, more earth-friendly ways, why bother to undertake the difficult and drastic changes that environmentalists call for? In fact, one of the tactics of anti-environmentalists in the "wise use" movement is to refuse the distinction between human and non-human change to the environment.[10] In this line of reasoning, species extinctions have been happening as long as there have been species, and the earth carries on regardless, so why suppose that the extinctions caused by humans are

10. See chapter 2, note 92.

different or worse? Every animal alters its environment for the sake of survival, comfort, or procreation. Humans are simply filling their particular niche in the natural order by their construction of dwellings, highways, power plants, and so forth.

In my claim that memories of a "green age" are unreliable, and that neither human society nor "pure" nature has ever, or will ever, match the eschaton, am I blurring the distinction between human sin and natural evil in a way that supports this "wise use" reasoning? I certainly hope not.[11] The existence of fatal disease, for example, has never lessened the reprehensibility of murder. My point is simply that neither disease, nor murder, nor other afflictions are God's ultimate intention for human and non-human creation.[12] The absence of suffering from the eschaton does not entail, however, the denial of suffering in the present world, or the erasure of its effects. Here we return to the role of memory in eschatological imagination.

11. The distinction between natural evil and sin comes under legitimate pressure from several directions. First, it depends upon a moral gap between humans, who choose sinfully, and animals, which are driven by instinct. But this gap lessens with each new zoological study of those animals whose minds are more complex and whose lives are more culturally shaped than has been supposed. If different orangutan communities, for instance, display different ways to crack open fruits, they might also display different ways of treating their mates or raising their young—behaviors that, when "chosen" rather than instinctual, are likely to be called moral. And on the other side, human "choice" is always shaped by unchosen factors—heredity, social conditioning, perception of available options, and so forth. See MacIntyre, *Dependent Rational Animals*. The second factor pressing against the natural evil-human sin distinction is the recognition that describing a disaster (or boon) as natural can mask the unjust political distribution of its effects. Floods and hurricanes along the Atlantic Coast, for instance, consistently hit poor communities harder because they occupy less desirable land in the flood plain. The answer is not to jettison the distinction, but to notice what work it does in an argument and what status it is given. My own inclination is to keep all such distinctions provisional and cautious, except for the distinction between Creator and creatures, the solidity of which relativizes all others.

12. Richard Fern holds that predation is part of God's plan for creation, because the faculties that make predators good at hunting are integral to their nature, and because conflict and struggle are necessary for higher-order goods (including love). However, he focuses too much on particular eschatological passages in the Bible and neglects the larger story of God's providence for the world. See Fern, *Nature*.

Rowan Williams, writing about Jesus' post-resurrection appearances, argues that memory is part of the identity of an individual, a community, a nation:

> I am what I am because of what I have been and done, good and bad. My self is woven out of a great web of complicated motivation, reflections, intentions, and actions . . . . And [we] need to be able . . . to take responsibility for some things and to accept the inevitability of others—to *own* the whole of ourselves, to acknowledge realities both past and present, to destroy all the crippling illusions about ourselves that lock us up in selfish fantasies about out power or independence. I depend on the past, and it is part of me; to deny it is to deny myself. I *am* my history.[13]

Not only is the past part of my identity, but also it is part of the history of God's faithfulness and activity in the world—however invisible it seemed at the time. For the prophets of Israel, therefore, the new age "means going back to the ruins of the past, to the devastated and depopulated land of Israel and building there, with the help of God, a city which is new but which still stands on the same earth as the old."[14]

Moreover, the resurrected Jesus still bears the wounds of crucifixion when the apostles meet him: the new age redeems, but does not deny, the wounds of the old. So the human self that receives forgiveness and redemption, that is taken up in God's life through Jesus Christ is the *whole* self, including not only the pretty past, but the ugly past as well.[15]

Williams goes on to point out that in the appearance stories, Jesus meets the disciples where he met them before: in the upper room, in Galilee, on the hill, on the seashore.[16] All the sites of terror and betrayal

13. Rowan Williams, *Ray of Darkness*, 49.

14. Ibid., 65. This is an important counter to the interpretation of the book of Revelation to mean that the current biophysical universe will be completely eradicated, and the new creation will bear no continuity with the old. See Russell, "*New Heavens and New Earth*," and Polkinghorne, *God of Hope*.

15. This has an expansive rather than a delimiting effect, in contrast to the old debates over what gets saved through Jesus, and the more recent insistence by some conservatives that because Jesus was male, only males can serve as priests.

16. Rowan Williams, *Ray of Darkness*, 65.

become sites of reconciliation and repair. "The resurrection is a recapitulation, not a reversal, of the history of Jesus *and* of his disciples. Their failures, his cross, all bound together, are memories never to be obliterated, but now taken up to be healed in the new age."[17] Jesus deliberately and physically confronts the disciples with the places and events of their shameful past, only to heal the places and memories by revisiting them, and thereby granting the disciples and their past a new destiny in his love.

Roberto Goizueta argues even more sharply that not only did the resurrected Jesus still bear visible wounds, but also "the wounds themselves became the instruments of Jesus' reconciliation with his friends. "Put your finger here. . . ." (Luke 24:39; and John 20:27).[18] Is it possible, he asks, that by asking the disciples to touch his wounds, Jesus

> was inviting them not only to believe in the truth of the resurrection but also, concomitantly, to acknowledge, confess, and repent of their own complicity in his crucifixion? Is it possible that the visible wounds engendered belief not only because they allowed the apostles to see the truth about the resurrected Jesus Christ but because at the same time the wounds allowed the apostles to see the truth about themselves?[19]

Denial of Christ's wounds is denial of our own culpability. For Goizueta, writing about Hispanic-American theology and mestizo history, denying the wounds is an all-too-common tendency in modern (white) theology, shaped by the mythology of progress. Turning memory into nostalgia means "the atrocities of war and conquest are transformed into tales of adventure, genocide is romanticized at the Alamo, and the period of slavery is wistfully recalled in the idyllic images of "Gone with the Wind."[20] But for Christians, the past, present, and future are intertwined: "Christ

---

17. Ibid., 66. Williams shows the explicitness of Jesus' recapitulation: "Peter's threefold denial is countered with his threefold confession and the Lord's threefold charge to him. And . . . there is a 'fire of coals' burning on the shore, just as there was in the High Priest's courtyard. . . ."

18. Goizueta, "Why Are You Frightened," 51.

19. Ibid., 52.

20. Metz, *Faith in History*, 109, quoted in Goizueta, "Why Are You Frightened?" 53.

has died, Christ is risen, Christ will come again." Eschatology cannot be legitimately "cleansed" of oppression, sin, or culpability in their particular historic manifestations. Christ's suffering on the cross is taken up into his glorified life, rather than forgotten.

It is in this way that our memory, collective and individual, shapes what we are able to hope for. A nostalgia scrubbed clean of sin and misery truncates the possibilities available to the eschatological imagination in two ways. First, we cannot begin to imagine the verdant fullness of the Kingdom without first seeing clearly the barren, bloody patches of our lives in a fallen world. We cannot imagine, for instance, true reconciliation between whites and African-Americans and Hispanics unless we acknowledge the failure to reconcile in the past and, more, the dreadful cost of that failure. We cannot imagine the lushness of New Jerusalem unless we recognize the toxic barrens that our cities have become. Second, we diminish the sovereignty of Christ's lordship if we limit the scope of his reconciling power. As the prophets of Israel recognized, *everything* is under Christ's care and everything has a future in Christ. In Williams's words, "risen life in and with Christ is *now*, entirely fresh, full of what we could never have foreseen or planned, yet it is built from the bricks and mortar, messy and unlovely, of our past . . . *Our* earth, *our* dull and stained lives, these are the living stones of God's New Jerusalem."[21]

Now what has all this to do with ecology? One way to begin to understand the connection is to recall the way the land figured in the identity and hopes of ancient Israel. Land—a particular geographical and historical place—was part of God's promise to Abraham. It suffered the effects of Israel's unfaithfulness, flourished under God's merciful care, and figured as a central element of Israel's messianic hopes.[22] Israel's identity, then, was bound up with her remembered past *as it took place on*

---

21. Rowan Williams, *Ray of Darkness*, 65. This is not to say that we should not imagine beyond the bounds of 'normal' life. Indeed, eschatology is exactly the attempt to understand, faithfully and imaginatively, what God promises for the renewal and utter transformation of this world. Typically, however, what we view as normal masks considerable ugliness and failure. The inability to see beyond the normal, therefore, results from a twofold denial that the ugliness is present and that God can and will overcome it.

22. See Brueggemann, *Land,* and Northcott, *Environment.*

or in relation to that land. The Israelites understood that human history does not float along in some a-topic realm; rather, people live and sin and love and die in particular earthly places. The land itself was part of what God would redeem, and likewise, for the topography of Jesus' ministry, betrayal, and crucifixion. In meeting the disciples there again, Jesus heals the brokenness of those people and that earth.

Places of human sinfulness are not, of course, the only parts of non-human nature to be redeemed. They do demonstrate, however, the inter-connection between human and non-human stories. Just as the human past and present, which shape our hope, need to be seen in full and free of delusions of innocence, so too the non-human past and present. It is important to confess both the length and breadth of humans' abuse of the earth as well as the imperfection of the cosmos independent of human activity. Environmental eschatology can be neither a romantic return to pre-Industrial harmony, nor a sort of debriding of earth's surface, scraping off humanity's destruction to reveal pristine wilderness. The present earth whose redemption we pray and hope for in the Kingdom, whose future we imagine in the eschaton, must be an earth full of pain as well as joy, waste as well as fertility, death as well as life. Otherwise, we deny the suffering of the very creatures we claim to honor, and devalue the sovereignty of the risen Christ.

This is not to claim that our image of non-human nature can—or should—be "objective" in a positivist sense or total in some omniscient sense. Certainly memories, images, or constructs of nature vary with culture, geography, history, and so forth. The story of nature is many different stories. What we aim for, then, is a "clear" image in two ways. First, we aim for freedom from sinful self-delusion, just as we do when we present ourselves before God in confession. We try to acknowledge our complicity in the destruction of ecosystems, in a flush-and-forget lifestyle, and in the shifting of toxic effects to those who are poorer or less powerful than ourselves. We also try to face the brutal aspects of nature as well as the beautiful and peaceful aspects. Second, we remind ourselves that our memory or story of nature is necessarily partial and provisional, certainly distorted by self-interest, and maybe wrong.

THE CHARACTER OF THE KINGDOM OF GOD

⁊ The task, then, is to read as best we can "our" story of nature and ourselves through the churches' narrative of the promised second coming and the eschaton. Given the caveats that knowledge of the Kingdom of God requires practice, and that human understanding will never match the divine intention, what do the Scriptures tell us about the Kingdom? In other words, what does "salvation" mean? Put this way, it is clear that the answer cannot be determined by reading only the "eschatological" texts. [23] (Likewise, Christian ethics cannot be defined only from the "ethical" texts.)[24] That is why this chapter is after the one on God's creating freedom and activity; the whole creation/redemption story is required for an adequate grasp of the eschatological vision. Furthermore, because the Reign is opposed to the anti-Reign (to use Sobrino's words again), and the anti-Reign is not singular, but varies according to historical circumstances, different aspects of God's Kingdom shine most brightly to Christians in different situations. It should be clear this does *not* mean the Kingdom of God is whatever Christians would like it to be. As Richard Bauckham writes, "the eschatological future of the world is not any conceivable future, but specifically that future which God has opened up for the world in raising the crucified Jesus to new life."[25] The reverse aspect of the Christological center, on the other hand, is that the only limits to the Kingdom are those set by God.[26] It is important that God's possibilities are infinite and unbounded by history, while human visions of the Kingdom of God are invariably constrained by the context of finite, earthly life. In the matter of context, it is all too easy to see the contextual limits to the other person's vision while remaining blind to our own. It is also easy to view contextual effects as negative, as flaws of an eschatology,

23. Katherine Sakenfeld, for instance, makes a good case for the story of Ruth as an eschatological vision. See Sakenfeld, "Ruth 4."

24. Sakenfeld, 56. This is part of the difficulty with Glen Stassen and David Gushee's *Kingdom Ethics* book. They rely so exclusively on the Sermon on the Mount that it sometimes seems abstracted from the biblical narrative. Stassen and Gushee, *Kingdom Ethics*.

25. Bauckham, "Eschatology," 209.

26. See the section on God's creative freedom in chapter 2.

rather than a Spirit-led response to the evils of the world. Yet, it would be more surprising (and less faithful), in a world where children starve, for a vision of the Kingdom *not* to include abundant food and joyful play.[27]

Therefore, as James Cone demonstrates, the songs of African slaves in the United States emphasized the reversal of current conditions. Slaves will be free; white people will not be in power, sins will be eradicated, and material existence will be comfortable and secure:

> No more hard trial in de Kingdom;
> No more tribulation, no more parting,
> No more quarreling, backbiting in de Kingdom,
> No more sunshine fer to burn you
> No more rain for to wet you.
> Every day will be Sunday in heaven.

As this spiritual indicates, heaven was not only free from human sin and guilt, but also free from difficult natural conditions as well. "Heaven was an affirmation of this hope in the absolute power of God's righteousness" over *all* things, a power that would be fully revealed at the end of history.[28]

> Dere's no whips a-crackin'
> Dere's no stormy weather.
> No more slavery in de Kingdom
> No evil-doers in de Kingdom
> All is gladness in de Kingdom!

In contrast, pro-slavery Presbyterians believed, as part of the myth of progress, that the existing social order was maturing into the millennium. They said, accordingly, that while slavery might not continue past the

---

27. This is a subtle point. On the one hand, I am not arguing that an adequate eschatology is *whatever* would reverse earthly sin and suffering; that would abandon the christocentrism of redemption. On the other hand, Jesus Christ highlighted, in his preaching and healing, real incidents of suffering and evil. So, just as the lives of disciples must serve those in need, so the eschatological understandings of disciples must serve those in need. As liberation theologians have pointed out, every theology serves the interests of somebody.

28. Cone, *Spirituals*, 93.

second Advent of Christ, it would certainly continue until then, as part of God's development plan for improving humanity.[29]

So far, this chapter has argued that knowledge of God's Kingdom is neither an exercise in cognitive theorizing nor a free-floating human construction of the "good life." It is, rather, the knowledge gained by seeing God's work, testifying—through word and action—to God's promise in particular historical situations, and remembering God's gracious presence in the midst of sin and evil. Christian visions of the Kingdom include the brokenness of past and present taken up and redeemed in the life of Christ. When that happens, the Kingdom is consummated, and the promises fulfilled. It is appropriate, then, to look now at the particulars of God's promises in the Old and New Testaments.

Not all visions of the eschaton reflect their historical conditions as vividly as the slavery examples. Most of them, though, contain a set of familiar elements: peace, justice, abundance, righteousness, communion with God, reconciliation, and love. The psalmist writes,

> Surely his salvation is at hand for those who fear him,
>   that his glory may dwell in our land.
> Steadfast love and faithfulness will meet;
>   righteousness and peace will kiss each other.
> Faithfulness will spring up from the ground,
>   and righteousness will look down from the sky.
> The Lord will give what is good,
>   and our land will yield its increase.
> Righteousness will go before him,
>   and will make a path for his steps. (Ps 85:9–13)

All of these characteristics pertain not only to humans, but also to *all* of creation.[30] Belief in the redemption of all creation has been part of the Christian tradition from the beginning. Although it has not

---

29. Maddex, "Proslavery Millennialism," 46–62.

30. Four-year-old Eli asked, "Mama, when will we see the dinosaurs?"

"Well, honey, you know they're all gone. We won't see the dinosaurs until we get to heaven."

"I *know*! But *what day* in heaven?"

always been the predominant view, it has never disappeared, even in times when a purely spiritualized notion of the eschaton seemed to prevail.[31] The following section describes the different key aspects of the Kingdom as expressed by the prophets and testified to by Jesus. It is necessarily brief and sketchy; the goal here is to recall the scriptural character of the Kingdom, in order to indicate what Christian disciples witness to when they witness to God's Kingdom.

PEACE

As Yoder writes, "concern for peace . . . is part of the purpose of God for all eternity."[32] The world was created, after all, not through a battle between gods, but out of the irenic, extravagant love of Father, Son, and Holy Spirit. Because the trinitarian God is prior to—before, after, beyond—the created world, peace *is* the ultimate reality. Violence and warfare (nature red in tooth and claw) may seem like the way of the world, but Christians believe that they are results of the Fall—contrary to God's intentions and destined for defeat. Christians themselves have not been very peaceful over the centuries, and apocalyptic fervor sometimes envisions the end of the world in spectacularly violent scenarios. Nonetheless, the Kingdom itself—after the dust of Armageddon settles—is always a realm of peace. Isaiah 60:17–18 is one of numerous prophetic texts that depicts that realm: "I will appoint Peace as your overseer, and Righteousness as your taskmaster. Violence shall no more be heard in your land, devastation or destruction within your borders." Jesus' incarnation fulfilled this divine promise. Jesus preached peace and demonstrated peace in his refusal to take up arms against his accusers and in his fellowship with outcasts. Above all, in his death and resurrection, Christ conquered war with peace and death with life.

The peace of Christ, however, is neither nuclear deterrence nor complacency. It is not a smooth surface laid over violent tensions un-

---

31. There is no comprehensive treatment of the Christian belief in the redemption of all creation. Partial treatments can be found in Galloway, *Cosmic Christ,* and Santmire, *Travail of Nature.*

32. Yoder, *He Came Preaching Peace,* 34.

derneath. The peace of Christ, rather, is the unity of Christ: all things (*ta panta*) united in their source, savior, and destiny. It is the result of reconciling differences and healing ruptures, of divine justice and mercy. This peace is not only interpersonal, but interspecies: not the wary tolerance of animals forced to share a waterhole, but shared communion with God. Richard Bauckham argues persuasively that the vision of Isaiah 11:7–9 is borne out by Mark 1:13 in which Jesus is "with the wild beasts" during his temptation in the wilderness.

> The expression "to be with someone" (*einai meta tinos*) frequently has the strongly positive sense of close association or friendship or agreement or assistance . . . Just as he resists Satan not as merely an individual righteous person but as the messianic Son of God on behalf of and for the sake of others, so he establishes, representatively, the messianic peace with wild animals.[33]

JUSTICE/LIBERATION/RECONCILIATION

Because God's peace is essential rather than superficial, it results from divine justice and liberation. According to the Old Testament prophets, divine justice "does not consist primordially in pronouncing an impartial verdict, but in the protection given to the poor, to widows and orphans."[34] Rather than "equal distribution" to an already unequal world, God's justice re-creates the world in accordance with mercy and righteousness. Stassen and Gushee note four dimensions to this understanding of justice: (1) deliverance of the poor and powerless from the injustice that they regularly experience; (2) lifting the foot of domineering power off the neck of the dominated and oppressed; (3) stopping the violence and establishing peace; (4) and restoring the outcasts, the excluded, the Gentiles, the exiles, and the refugees to community.[35] Jesus Christ, who preached the Kingdom as a radical upheaval of the fallen earthly order, embodies the partiality of divine justice. Sobrino notes, then, that

33. Bauckham, "Jesus and the Animals II," 58.
34. Sobrino, *Jesus the Liberator*, 83.
35. Stassen and Gushee, *Kingdom Ethics*, 349.

when Christ is identified simply as "love," this identification may hide the prophetic forms of Jesus' love, "such as justice and loving partiality for the poor."[36]

Divine justice, however, is not content with turning losers into winners, but in reconciling all conflicting parties. Scott Bader-Saye notes that for Jesus "to "justify" is to make right, to reconcile, to bring back a proper ordering of human life and relationships."[37] And, I contend, this biblical vision of reconciliation includes non-human creation. Relationships between lands and their animal inhabitants, between human communities and their non-human neighbors, all are reconciled into the unity of Christ.

ABUNDANCE

Material abundance tends to get little attention in most analyses of Kingdom texts (except as wealth). It seems too obvious to mention, perhaps, or too expected. Of course there will be sufficient food, water, and land in the Reign of God! All the references to heaven as a "banquet" or feast, the miracles of loaves and fishes, even the disciples' habit of journeying with few provisions—all these point to the Kingdom as the "land of milk and honey."[38] To the millions who are starving, however, or whose children suffer continual hunger, the prospect of abundant food is indeed a matter of hope, rather than expectation. During a meeting with Pope John Paul II, a Salvadoran representative pleaded, "We are hungry for God. And we are hungry for bread."[39] Nearly two-thirds of the world's people live where clean water, food, or arable land is insufficient, either through "natural" or social causes, or a combination of both. Their prayer for daily bread is heartfelt. Moreover, material abundance is much more

36. Sobrino, *Jesus*, 15. Note that Sallie McFague and Andrew Linzey have used the partiality of God's love to develop an argument for improving human treatment of animals. See McFague, *The Body of God*, and Linzey, *Animal Theology*.

37. Bader-Saye, "Aristotle or Abraham," 540.

38. Isa 25:6; Matt 8, 10:8–9, 22; Luke 14. In the Gospel of John, Jesus himself is the feast. John 7:37.

39. Holy See press office. "Synod Episcoporum Bulletin," September 30-October 27, 2001.

subversive, even revolutionary, than it might appear. Modern Western capitalism is premised on the assumption of enduring scarcity—that, at bottom, there are not enough resources to satisfy the desires of all, because "unlimited desire is a natural quality" of humans.[40] So competition is the norm; the market adjudicates between supply and demand; and affluent people are encouraged to amass wealth for "security" against the possibility of hardship. Taking more than we need (if we have the money to do so) becomes, on this view, a mark of prudence rather than greed. There are winners and losers, and the goal is to be on the winning side. Contrast this to the New Jerusalem where the water of life flows through the city and is given without cost to all who desire it.[41] Contrast scarcity with the vision in Amos 9:13–15, where "the mountains shall drip sweet wine, and all the hills shall flow with it."[42] Eschatological abundance is not only a future vision, but also a reality prefigured in the Eucharist, where the body and blood of Christ are shared with all.

Abundance in this respect means material plenty, enough for all. It does *not* mean, however, that human greed will finally be satisfied. As Sobrino says, "if the term abundance describes a source of blessing in the Old Testament, the term 'riches' implies a motive for cursing in the New Testament."[43] The Kingdom of God is not a lottery prize, or a giant shopping mall with the prices marked out. Greed and gluttony are forgiven but also eliminated, supplanted by generosity and caritas. Hence the Gospel stories of Jesus castigating those who are rich or those who do not share, and the sharing of goods described in Acts 4:32–35.

Note too, that eschatological abundance is not limited to human lives. Animals, too, receive ample food and water, and comfortable surroundings.[44] Irenaeus makes a beautiful comment on the fact that predators (including humans) will be vegetarians in the Kingdom: ". . . that the lion shall [then] feed on straw. And this indicates the large size

---

40. Achterhuis, "Scarcity and Sustainability," 106.

41. Rev 21:6; 22:17.

42. See also Ezek 36; Zech 9:15.

43. Sobrino, *Jesus*, 171.

44. Joel 2:22; Isa 11:7–9; Ezek 47:9–12.

and rich quality of the fruits. For if that animal, the lion, feeds upon straw [in the eschaton], of what a quality must the wheat itself be whose straw shall serve as suitable food for lions?"[45] To think of a world that is infinitely fruitful strains many modern sensibilities.[46] Recall, however, the Christian insistence on God's freedom in creation (chapter 2). This world and all its arrangements are both contingent on God's providence and fallen from God's intention; it is part of Christian faith that God's reign will establish fecundity and plenty for all creation.

RIGHTEOUSNESS

Just as peace is not merely the "temporary cessation of hostilities," so righteousness is not merely the restraint of sinfulness. The Bible often pairs righteousness with justice: righteousness delivers justice that releases the oppressed and restores community.[47] But biblical righteousness is a wide-ranging ideal, including right relationships with God and neighbor, and faithfulness and consistency in all areas of life. Righteousness, therefore, cannot be limited to human-human or human-divine interaction. In fact, the very idea of salvation as restoration of right relationships has ecological overtones.[48] If all things on heaven and earth will be united with God through Jesus Christ, then right relationship—mutuality, fellowship,

---

45. Irenaeus *Against Heresies* V.33.4. Note, too, that since the prohibition against eating flesh was not lifted until after the Flood, the animals on Noah's ark would have all been vegetarian.

46. Even Richard Clifford, a biblical scholar, says, "no one with a grain of sense believes that . . . Isaiah 11 is intended literally, as though the digestive system of a carnivore were going to be transformed into a herbivore." I cannot imagine why a God who incarnated and resurrected Jesus Christ should be inconvenienced by biology. Nor do I see why Clifford, a self-identified Christian, should take such a scornful attitude on this point. Clifford, "Bible and the Environment," 9.

47. Stassen and Gushee, *Kingdom Ethics*, 42.

48. LaCugna, *God for Us*, 284, 292.

union without homogeneity—with all God's creation, beyond our current imaginings, will comprise the very "atmosphere" of the eschaton.[49]

COMMUNION WITH GOD

Eternal communion with God is not another attribute of the Kingdom; it *is* the Kingdom. This is often described in the Old Testament as God's glory or light shining over all, or God's presence with his people in an intimate fashion, far beyond anything possible in this world (e.g., Isa 9:2; 30:26; 32:15; 42:16; 53:11; 60:20). New Testament writers point to Jesus Christ himself as the "Eschatos," the Last One, Alpha and Omega, the One who comes to us. Jesus proclaims, "Fear not, I am the first and the last, and the living one; I died, and behold I am alive for evermore, and I have the keys of Death and Hades" (Rev 1:17–18). In the reign of God, Christ's lordship over all things puts all things under subjection to God, and draws all creation into the loving communion between Father, Son, and Spirit (1 Cor 15:28). This intimate contact with God is promised, yet mysterious; it has been described as seeing God clearly (1 Cor 13:12; 1 John 3:2), as theosis (Athanasius, Irenaeus, Augustine, Maximus), as

---

49. The emphasis on right relationship as a quality of the Kingdom and on relationality as central to Christian understanding of the Trinity (*à la* Elizabeth Johnson, Christof Schwöbel, and Catherine LaCugna) triggers a concern. Does it risk trivializing matter once again? For instance, does it matter if tigers get redeemed as long as we can have tiger-ish relationships with another? The response, I suspect, is that even in the eschaton, perhaps especially in the eschaton, tiger-ish relationships involve real tigers. The point of the emphasis on relationality is not to make bodies disappear, but to recognize that bodies always already live within relationships to others. This is reminiscent of the last verse of Yeats' "Among School Children":

> Labour is blossoming or dancing where
> The body is not bruised to pleasure soul.
> Nor beauty born out of its own despair,
> Nor blear-eyed wisdom out of midnight oil.
> O chestnut-tree, great-rooted blossomer,
> Are you the leaf, the blossom or the bole?
> O body swayed to music, O brightening glance,
> How can we know the dancer from the dance?

becoming friends with God (Aquinas), and as abiding in God (Basil). No one can possibly know how this will be, but Christians believe they receive intimations through sacraments, moments of grace, and prayer.

Because communion is not union, and because the ultimate reality of the universe is Trinity, eschatological union with God does not entail removal of all distinctions.[50] Resurrected persons are in some way continuous with their earthly selves, and so each is unique. Likewise, there is no theological reason to suppose that nonhumans will lose their individuality.[51] The Kingdom is a realm of transformed selves, landscapes, and relationships, suffused with the light of God's presence.

Obviously, this is not an exhaustive description of the Kingdom; more characteristics could be added and these could be analyzed more fully. The characteristics on this list, however, are indeed central to any vision of the eschaton. Moreover, they serve to display the all-encompassing nature of the Kingdom, and the ways that Christians can understand the redemption of the non-human creation.

THE CHARACTER OF CHRISTIAN WITNESS

❧ At this point, having reviewed what the promises of God's Kingdom include, it is important to examine the character of witness itself. How can human lives, how can churches, proclaim the Kingdom to the present world? The Kingdom of God is a realm of liberation, reconciliation, justice, peace, and abundance, in the presence of the Triune God. The Kingdom of God will be consummated when the Son redeems the created world in God's own time. Christians are commanded to spread the news

50. LaCugna, *God For Us*, 299.

51. John Polkinghorne writes, "What are we to expect will be the destiny of non-human creatures? They must have their share in cosmic hope, but we scarcely need suppose that every dinosaur that ever lived . . . will each have its own individual eschatological future," and that too inclusive a resurrection belief puts the eschaton "in danger of becoming a museum collection of all that has ever been." This reluctance to include non-human creation fully in the eschaton contradicts much of the Christian tradition as well as the infinite caritas of God. The eschatological being of dinosaurs is no more a "museum collection" than the eschatological being of human persons, even though dinosaurs and humans may not participate in divinity in the same way. Polkinghorne, *God of Hope*.

of Jesus Christ who ushered in the Kingdom, and to follow the path of that Jesus through service to others, even if it leads to the cross. But these two are not separate commandments: it is through self-sacrificial service that Christian lives exemplify the truth of Christian teachings. So witness is both proclamation and demonstration, preaching and doing, much as Jesus proclaimed the Kingdom of Heaven and demonstrated it through his ministry. Thus witness always points beyond itself—it is made possible by the Spirit who sustains all things in their pilgrimage to God.

FAITHFUL OBEDIENCE

At this point it is important to recall that the fulfillment of the Kingdom, just like its inauguration through Jesus Christ, is absolutely and entirely a gift from God. This view prevails throughout the Old Testament: whatever they do, "mortals cannot abide in their pomp; they are like the animals that perish" (Ps 49:20). It is God's glory that will fill the earth, God's arm that will triumph. God does not, in the Old Testament, charge humans with the general fulfillment of God's plan. John Howard Yoder, in a discussion of Old Testament holy wars, writes, "one of the traits of the Old Testament story . . . is the identification of YHWH as the God who saves his people without their needing to act."[52] In particular, God's people are not to act in such a way that presumes their own control of events, their own accomplishment of the world's salvation. "Old Testament eschatology understands the future to be completely in the hands of God."[53] Witness, then, is an "ethic of *response*" to God's grace and love of the world that will be consummated in the Kingdom.[54] Just as the rest of creation, human activity gains meaning from an extrinsic source. Christian ethics, from this perspective, is less about figuring out what to do on the earth than about responding to what God has done and continues to do. The initiative is always with the God who creates, sustains, and redeems the world. Because this is the God Christians tes-

52. Yoder, *Politics of Jesus*, 76.

53. Gowan, *Eschatology*, 122.

54. This is Verhey's term used to describe Jesus' ethic and, by extension, Christians' own ethic. Verhey, *Great Reversal*.

tify to, Christian witness turns out to be the route to human flourishing; following Christ is the way to become truly human.[55]

The relationship between Christian witness and the eschaton is crucial; therefore, some ways of describing this relationship are theologically stronger than others. I do not claim that certain language is universally correct, but that some ways of characterizing this relationship are theologically and practically treacherous. It is difficult to capture the unlimited sovereignty of God without making earthly activity trivial, but this is what "witness" language attempts to do. "Building the Kingdom" is a common phrase among both Protestants and Roman Catholics.[56] It evokes a sense of upward movement, and measurable progress toward an obvious goal. "Building the Kingdom," however, alarms some theologians because it sounds as though the Kingdom is a *human* production rather than a divine gift. It seems to dodge the "eschatological reserve," the understanding that nothing humans accomplish can ever match the eschaton. The Kingdom of God relativizes every earthly relationship, social structure, or political system. It is not a project that awaits the right technology, or a better theory, or larger church membership, or even more obedient disciples, so it is hardly something that can be "built." For Christians on the "upper side" of history, those Westerners whose affluence, race, and power protects them from most of the evil and suffering in the world, "building the Kingdom" sounds too close to "managing the world" toward our own vision of a "new world order." And this is risky thinking indeed; as Schwöbel writes, "the temptation to rule is the

55. For reasons detailed in chapter 1, I have avoided the question of humanity's "role" in relation to non-human creation. I think we can make this anthropological claim without implying a precise answer to that question. Becoming human-qua-human requires witnessing to God through self-giving love. To know this does not require that we know what it takes to become elephant-qua-elephant or mountain-qua-mountain. Moreover, as Hütter explains, "the primary concern of a 'theology of creation' is not the creature but the *Creator*, since in God's activity alone are rooted the promise and the claim inherent in 'creation.'" Hütter, "*Creatio ex nihilo*," 90.

56. It appears in "For Us, the Kingdom is Grace," *L'Osservatore Romano* by John Paul II, December 6, 2000, although it contradicts most of the document's argument. See the Vatican's website at http://www.vatican.va/holy_father/john_paul_ii/audiences/2000/documents/hf_jp-ii_aud_20001206_en.html (Accessed March 23, 2005).

serpent's promise," even if the ruler is well-intentioned.[57] To regard the human task as "building the Kingdom" in this way is to fundamentally misunderstand the nature of the universe, God, and humanity. Drawing on Hendrikus Berkhof, Jim Wallis says, "we are not asked to defeat the powers. That is the work of Christ, which he has already done and will continue to do. Our task is to be witnesses and signs of Christ's victory by simply standing firmly in our faith and belief against the seduction and slavery of the powers."[58] We witness to the reality of Christ himself, and to the work he has already done, is presently doing, and will do in redeeming the world. When powerful people regard the Kingdom as a human (or even earthly) project, they tend toward either violence (*making* the world safe, or right, or just) or despair (the world will *never* be made right).

However, this argument does not apply to Christians on the "underside" of history. Dan Bell and Jon Sobrino demonstrate that the eschatological reserve is not an issue for Latin American theologians, who work with Christians under conditions of severe poverty, suffering, and oppression. Their world is obviously and dramatically *not* the Kingdom of God! In such a case, where people often feel both powerless and hopeless, the phrase "building the Kingdom" may convey a sense both of power and of hope. Sobrino notes that the very gratuity of the reign of God simultaneously calls forth and enables Christian praxis in response:

> Recall that Jesus proclaimed the gratuity of the Reign, and at the same time he himself exercised a practice and required one of others . . . what Christian faith does is proclaim where the initiative is, and what it means for practice that the initiative should be with God. It means that practice must be performed not with hubris but with gratitude; that God's first practice, the antecedent unconditional divine love, shows how historical practice is to be carried out and how one is enabled to perform it . . . In the most gratuitous of all the divine acts, God has stamped us with this analogy with the divinity, that we may be with others what

57. Schwöbel, "God, Creation and Community," 172.

58. Wallis, *Agenda for Biblical People*, 26. See Berkhof, *Christ and the Powers*, and Yoder, *Politics of Jesus*.

God has been with us, that we may do for others what God has done for us, and that we may deal with others as God has dealt with us.[59]

In other words, Christians' witness to the Kingdom is not simply watching, but pointing toward God's gracious creating and redeeming activity with the activity of their own lives. Yet, this sort of activity is itself laborious, costly, and discouraging, especially for Christians whose resources are depleted by poverty, hunger, disease, or repression. To speak of "building the Kingdom" in these circumstances is hardly a temptation toward domination, but an encouragement toward hopeful perseverance—given, of course, the cautions that Sobrino notes. As Bell implies, Christians in positions of stability and power tend to brandish the eschatological reserve against practice that opposes or subverts the existing social order, while ignoring its power to undercut that same order.[60] Neither the rulers nor the powerless can "build the Kingdom"; chances are, though, the rulers need reminding of this more often.

Another common way to describe earthly activity in relation to the Kingdom of God is that humans "participate" in the Kingdom in some way. "Participation" has an advantage over "building" in that it implies shared activity; one participates in praxis that is already occurring, that transcends the individual. It also resonates with the idea (mentioned above) that, in the eschaton, humans participate in the divine being of the Trinity. Hooker employs the idea of participation to describe both God's sustaining presence throughout creation and Christ's saving work in a fallen creation. But this is clearly a description of God-initiated

---

59. Sobrino, "Reign of God," 65.
60. Bell, "The Insurrectional Reserve," 643–75.

participation, different from a claim that we participate in God's work.[61] Look, for instance, at the way Stassen and Gushee write,

> New Testament scholars show that the Kingdom of God is not about what God does while humans stand by passively; nor is it about our effort to build the Kingdom while God passively watches. The Kingdom of God is Performative: it's God's performance in which we actively participate.[62]

The first sentence here is correct. The second, however, runs the risk (especially for Christians accustomed to managerial roles in society) of portraying humans as partners or helpers, whereas nothing in Scripture supports this view. The Kingdom is not something humans help to create or unfold, but something we receive, inherit, or enter. It precedes and transcends earthly life, and in no way depends upon human activity. The Kingdom, after all, is the salvation of the world, and humans are neither capable of, nor responsible for, enacting their salvation (or that of anything else).[63] So the following statement by Stassen and Gushee does not adequately guard against an anthropocentric reading:

---

61. Hooker, Book I.xi.5; Book III.ii.8; Book V.lvi.5. See also Allchin, *Participation in God*. Another possible, and more appropriate, use of participation language is to say that we participate eucharistically in the life of the Trinity and in the eschaton. Thus Cavanaugh writes about the Eucharist, "The form of a meal invites the participants to a physical, not merely spiritual, communion. Heaven and earth are united in the eucharistic meal, which anticipates the resurrection not merely of the soul but of a glorified version of the same body which now feasts on earth." This use of participation language signifies the eschatological union of earth with heaven that is brought about through God's action in the sacrament: humans participate in the divine reality. This is quite different from claiming that humans participate in the creating of that reality. Cavanaugh, *Torture and Eucharist*, 224.

62. Stassen and Gushee, *Kingdom Ethics*, 20.

63. Stassen and Gushee are not isolated examples; in fact, they are more careful than many other scholars. See, for example, Baker-Fletcher, *Sisters of Dust*; Isasi-Díaz, "Mujerista Theology's Method"; and especially O'Loughlin, "Ecotheology and Eschatology." On the other hand, Cone, Haas, Hopkins, and Copeland are conscientious in preserving the sense of God's initiative and sovereignty. Cone, *Black Theology of Liberation*; Haas, "Significance of Eschatology," 325–41; Hopkins, *Heart and Head*; Copeland, "Journeying," 26–46.

the good news is that the Kingdom has dawned in Jesus, and its final triumph is inevitable. Yet *coworkers* are needed. A Christian is one who humbles himself and chooses to enter in discipleship, to follow Jesus' path, to build his life upon his teachings and his practices even at great cost, to pass those teachings and practices on to others, and thus to enjoy the unspeakable privilege of *participating in the advance of God's reign* . . . Disciples of Jesus Christ both taste the joy of Kingdom living and are used by Jesus to *advance* the Kingdom until he comes again.[64]

The second risk of "participation" language is the idea that the Kingdom is arrived at in a gradual forward movement, rather than an apocalyptic upheaval. The Kingdom is viewed as the end of history, its goal in a temporal sense. Yet, the Kingdom is not the "last station" on the track of history; instead, it is the complete redemption of the whole track—first, last, middle. We do not and cannot know the relation of current history to the eschatological event—whether it will come at a bad time, a good time, or whatever. This ignorance condemns both the apocalyptic fervor in parts of the religio-political right and the progressive ideology of the left. It also means that, while human activity may be in harmony or disharmony with the nature of the Kingdom and the will of God, we neither nudge it forward nor hold it back. McCabe explains it this way:

> Should we think of the second coming as the culmination of the work of Christians in the world gradually leading up to this fulfillment; or should we think of it as something quite unexpected, as coming like a thief in the night . . . the second alternative is the one that finds most support in the New Testament. The difficulty that Christians find today with this is that it suggests that any "preparations" for the coming of Christ are an irrelevance; it suggests that Christians have really no

---

64. Stassen and Gushee, *Kingdom Ethics*, 30, my emphasis. The authors cite the great commission (Matt 28:19–20) as support for this claim. Yet, evangelism, proclaiming the good news and making disciples, is not the same as establishing the Kingdom; their inference seems unjustified. Even if every person on earth were Christian, the world would still be fallen, awaiting redemption by Jesus Christ.

political role in the world, it is useless for them to try to transform society, they ought instead to retire into themselves . . . [65]

What is at stake in the language issue is the essential and constant awareness of our creaturely dependence upon God. Without that awareness, not only are we tempted to force our own version of the Kingdom onto the world, with disastrous results, but we also reduce Christianity to little more than an inspirational story.[66]

I am not rejecting participation language altogether; in faithful community, with the Spirit's help, Christians do participate in the Kingdom of God; this is part of Jesus' message to the world. Christians do not, however, knowingly participate in *bringing forth* the Kingdom. Rather, what Christians do is *witness* or testify to the reality and promises of the Kingdom through Jesus Christ. Jon Sobrino's work may help clarify what I mean by witness, and also why my refusal of human power to bring about the Kingdom does not mean that human activity is not meaningful or important.

Sobrino begins by asking, "if Jesus thought that the Reign of God was imminent and gratuitous, then why might he not have restricted himself to its proclamation? Why not await that coming in passivity and confidence? Why not accept the situation of his world, if it was soon to change?"[67] In fact, "Jesus performed a series of activities that he understood as *signs of the Reign*. As signs, they are not the totality of the Reign."[68] Yet, they make the Reign present and express the character of the Reign. In Jesus' ministry, that is, he performed miracles, welcomed sinners, and expelled demons, as well as preached the Kingdom of God. The miracles were signs of salvation from illness, hunger, affliction, and death. In welcoming sinners, Jesus posited signs of the inclusiveness of the Kingdom and its "upside-down" character; those whom society

---

65. McCabe, *What is Ethics All About*, 159.

66. Other phrases used in relation to the Kingdom include "living into the Kingdom," "pilgrimage toward the Kingdom," and so forth. Each has its advantages and disadvantages, and each must be evaluated along the lines sketched here.

67. Sobrino, "Central Position," 48.

68. Ibid., 49.

despised would be welcomed into the Kingdom. In expelling demons and confronting his opponents, Jesus denounced the anti-Reign.[69] These are the same activities Jesus demands of his followers: relieving the suffering of others, welcoming the outcast, and denouncing the powers of injustice, hypocrisy, and oppression. "The follower is a witness, someone who reproduces—in historicized fashion—the life of Jesus . . . ."[70] In following the path of Jesus, the disciples themselves posit signs of the Kingdom of God. These signs are not the Kingdom itself; healing one person does not eradicate disease, feeding the crowd with a few fish does not eliminate hunger. Nonetheless, these signs point to the new *possibility* that disease and hunger *will* be eliminated. What is possible is not limited to what has happened before; reality itself has broken open to reveal, here and there, the culmination of God's Reign.

The coming of the Kingdom, therefore, "demands a conversion, metanoia, which is a task for the listener: the hope the poor must come to feel, the radical change of conduct required of the oppressors, the demands made on all to live a life worthy of the Kingdom."[71] The disciples—all Christians—are not bound by the limitations of history, because God has burst through those limits and boundaries. Even death cannot muffle Christians' proclamation of the Kingdom, because Jesus' resurrection has conquered death through loving obedience. The limits of discipleship, therefore, are the strictures of obedience to divine command.[72] Christians witness to the eschaton by living in accordance with

---

69. Ibid., 49–53.

70. Sobrino, "Spirituality," 238.

71. Sobrino, *Jesus*, 76.

72. Herbert McCabe writes about the Ten Commandments on Mount Sinai: "It must seem a strange thing that the first thing the God of freedom should do is to lay commands and restrictions on his people—and such absolute and uncompromising commands. You might expect the God of freedom to leave men free to make their own decisions and run their own lives. Freedom, we might expect, would consist in people doing what they liked. As a matter of theory this is no doubt true but there are various complications attending the notion of 'doing what you want.' As a matter of history one of the peculiar things about man is that when he is left to do exactly what he likes he straight away looks around for someone to enslave himself to, and if he can't find a master nearby he'll invent one." McCabe, *What Is Ethics All About?*, 115.

God's will; that is, by serving all creatures in their Spirit-led pilgrimages to God's Kingdom. John Booty, following Hooker, phrases this well:

> With power Christ created the world, but restored it by obedience . . . After the ascension power is defined by humility. True power is not only in regard to people but to the whole world, the animate and the inanimate, the biosphere and particles of energy. This is cause for rejoicing because we are now, in Christ, free to be obedient to the law of the universe, the law of love toward God and neighbor.[73]

Being freed of the constraints of "precedent" as well as sinfulness means that disciples are also freed from the responsibility to control the outcome of their actions—which creatures cannot do anyway. With regard to ecological action, Conyers notes,

> even if we affect a small part of the earth, our actions are fraught with all kinds of ambiguity. Our intentions to do a good thing always invite the possibility of unintended evil. We work hard to enrich human life and, destroying the environment, threaten the possibility of life itself . . . We are everywhere faced with the sobering realization that we cannot create the world, or even re-create it; we can only respond to what God has given.[74]

Instead of trying to reason via *means-end effectiveness*, Christians understand that the end—not of their actions, but of the world—is already guaranteed by Christ. The standard for Christian action, therefore, is not efficacy, but faithfulness.[75] Thus Ezekiel spoke the words given him by the Lord to the house of Israel, even though it was clear the Israelites would not listen. And those German Christians who resisted the Nazis, like Lutheran Pastor Franz Jägerstätter, sacrificed lives not because they believed their actions would stop the Nazis, but because they were faith-

---

73. Booty, *Reflections on Hooker*, 54.

74. Conyers, "Living Under Vacant Skies," 14.

75. Yoder, *Original Revolution*, 59.

ful to God.[76] This emphasis on obedience is not an invitation to stupidity, but a reminder that Christian prudence is sometimes the opposite of secular prudence (which usually equates with means-end thinking). For Christians, the "prudent thing to do" may include giving away one's money (regardless of the tax deduction), living among prostitutes and drug dealers, or standing between Israeli tanks and Palestinian rockthrowers. Prudence for Christians is the practical reasoning toward the action that best testifies to God's love for the world and promises of the world's salvation, even if the short-term prospect is dismal failure.

THE COSTLINESS OF WITNESS

Thus witness is nearly always *costly*, because, to use Sobrino's words, it always opposes the anti-Reign. The world we live in is a crucifying world.[77] Sometimes, for Christians in affluent circumstances, the cost is the difficulty of re-arranging one's life: using public transportation, turning off the air conditioning, giving more money to the church. Sometimes, especially for Christians in already difficult circumstances, the cost is liberty or even life. As Haas writes,

> the continuing presence of sin also requires Christians to embrace justice and obedience in the way of the cross, that is, with the understanding that it involves self-denial and the expectation of persecution and suffering for the sake of the Kingdom.[78]

This is why, so often, comfortable Christians dodge the call to witness; embracing justice and obedience entails resisting the overwhelming seductive powers of the anti-Reign in which we are already caught. With regard to the massive extermination of non-human species, for instance, William Greenway notes,

76. Before his execution for opposing Hitler, Jägerstätter wrote that although he "could change nothing in world affairs," he could at least "be a sign that not everyone let themselves be carried away with the tide." Catholic Peace Fellowship, "In Light of Eternity."

77. McCabe, *What Is Ethics All About?*, 132.

78. Haas, "Significance of Eschatology," 234.

many people deny the sacredness of life because it is too painful to acknowledge . . . It is easier to say "They're just animals," or "That's the way it is." But the Bible asks more of us . . . For the person who loves all creatures, life will become harder than it would be if one lived for oneself. But it will be richer and more beautiful as well.[79]

A life of witness is often harder than a life of surrender to the anti-Reign, but only through such witness and God's surprising grace can the Kingdom of God be glimpsed and finally received. The cruciform life is the only way to salvation.[80]

HOPE

What enables Christians to face the anti-Reign, to persevere through apparent failure and their own entanglement in sin, is their faith and hope in God's providence. Witness is always *hopeful*; it shares Christian hope as much as it shares material blessings. Hope is the essence of Christian eschatology as well as a central Christian virtue. Christian hope is not equivalent to optimism about human nature's eventual triumph over adversity, or about natural evolution toward the good. Instead, Christian hope is based on the resurrection, in which evil is overcome with the greatest of goods—divine love.[81] McCabe writes,

> the Christian moral outlook is essentially drawn from our contact with the future. It's based on the virtue of hope. It transcends

79. Greenway, "Animals and the Love of God," 681.

80. We probably cannot know in advance when Christians might be called to offer their lives for the sake of non-human creation and God's care for it. We do know that wildlife park rangers in Africa have been attacked by poachers, and that lumber companies in Mexico have murdered an activist who resisted their illegal logging. Perhaps a sign that U.S. Christians are engaging in serious discipleship will be that they, too, risk more than being ignored or scorned. See Catherine Bremer, "Mexico Murder Shows Grim Face of Illegal Logging," June 8, 2007, http://www.reuters.com/article/environmentNews/idUSN0837957320070608?feedType=RSS. See also Leon Marshall, "World's Park Rangers Seek Protection," September 5, 2003, http://news.nationalgeographic.com/news/2003/09/0905_030905_parkrangers.html.

81. Hare, "The Virtue of Hope," 23.

the present and is never wholly explicable in terms of the present because it is revolutionary. Thus the Christian moral position will always in the end seem unreasonable to the contemporary world.[82]

Hope does seem unreasonable, in our "cloven world" rife with suffering and loss.[83] This is why hope is a gift and a virtue rather than a mood or attitude—it does not rely on positive experiences, but on a reading of the world itself as trinitarian, as redeemable and redeemed into the love of the Triune God. The most cogent expressions of Christian hope may come from black and liberation theologians, for whom hope battles against everyday pressures of racism and poverty. Paraphrasing Howard Thurman, Genna Rae McNeil writes,

> no experience, no event at any particular moment in time or space exhausts what life is trying to do . . . To accept the gift of salvation is to live in the full knowledge and unassailable belief that after Jesus Christ's ministry, crucifixion, and resurrection, no human act can forever eradicate the possibility of possibility. To accept the gift of salvation is to live in the full knowledge and unassailable belief that after Jesus Christ's ministry, crucifixion, and resurrection, no human act is the last word on or work of love and justice.[84]

Yet, hope is not an invitation to wait passively for God's rescue vehicle. Sobrino shows that hope and praxis cannot be separated, for hope without praxis becomes abstracted from earthly reality. Old Testament prophets hoped for fulfillment of specific, material needs, and Jesus' miracles provided salvation from specific, material afflictions. Therefore,

> it is in practice that we learn what generates hope in the poor. Many good deeds can be done in behalf of the poor. These good deeds alleviate their needs. But not all good deeds, however

82. McCabe, *What Is Ethics All About?*, 154.

83. "Cloven world" is Joseph Sittler's wonderful way to describe how evil saturates the created world.

84. McNeil, "Waymaking and Responsibility," 59.

welcome, generate hope. It is in practice that one decides which signs, which proclamation of the Good News, which denunciation, which seedlings of a new society generate hope and therefore point in the direction of the Reign.[85]

The practice of faithful witness to the Kingdom requires, as well as generates, hope throughout the community. "If the acts of mercy don't arouse hope that it is possible for the Kingdom of God to come—not just that individual wants will be alleviated—and if they produce no sort of conflict, then they cannot be compared to the miracles of Jesus."[86]

Christian hope is future-oriented, but it encompasses the past as well. What Cone writes about black eschatology should be true of all Christian eschatology: "To grasp for the future of God is to know that those who die for freedom have not died in vain; they will see the Kingdom of God. This is precisely the meaning of our Lord's resurrection, and why we can fight against overwhelming odds."[87] Cone also points out that eschatological hope is not an impartial, disinterested hope, but a hope for God's liberation of the oppressed and judgment on the evildoers. It is easy for those with nothing to hope for a reversal of the present order. For Christians on the "upper" side of history, though, living into this hope requires the courage and skills to "unlearn their privilege"[88] that makes life in a fallen world relatively comfortable and easy. Such Christians must hope, not only for God to strip away excess possessions, status, and power, but also for God to cleanse them of their deep attachment to these things. All things and all people will be restored to their proper place in the Kingdom of God.

JOY

This is "good news," a matter of great *joy*, and Christian life testifies not only to the hope of the redemption of the world, but also to the joy of

---

85. Sobrino, "Central Position," 64.

86. Sobrino, *Jesus*, 92.

87. Cone, *Black Theology of Liberation*, 141.

88. This is Gayatri Spivak's phrase. See "Can the Subaltern Speak?" 66–111.

the Kingdom already present. Given the difficulty of witnessing against the anti-Reign, along with some cultural assumptions that holiness must be solemn, it is easy to neglect the joy. Yet, as Sittler points out, "The gravity of a life determined by God, lived to the glory of God, is not necessarily incongruent with abounding joy."[89] In particular, Christians should joyfully celebrate the grace and beauty of God's creation—fallen as it is—and the promise of Christ's redemption of the whole cosmos. "In Christ we are not only enabled to see the grace that inheres in the world as God's creation: God's action in Christ can give us the capacity to respond appropriately to creation-as-grace."[90] Joy, hope, and freedom together create what Sobrino calls "eschatological fullness" in the following of Jesus.

> Hope comes about against resignation, disenchantment, triviality; freedom comes about against the bonds that history imposes on love (risks, fears, selfishness); joy comes about against grief . . . so fullness becomes present not only against the not-yet of the limited but also against the certainly-not of oppression and dehumanization.[91]

Joy, hope, and freedom enable Christians to live in Christ as witnesses to Christ, with patience and perseverance.

WITNESS HAS NO PREDETERMINED LIMITS

Now, because Christ is Lord of all, Christian witness cannot on principle exclude any areas of earthly life. *Anything* from choosing toothpaste to choosing a career may become an area of witness. This does not mean that every decision or every aspect of life bears significant ethical weight, but that any of them *might* do so, depending on the work of Christ in the current historical situation. For instance, who would have expected, thirty years ago, that drinking particular types of coffee would become a way of witnessing *against* unjust trade practices and *for* the equitable distribution

89. Sittler, *Evocations*, 55.

90. Sittler, *Evocations*, 9. See also Booty, *Reflections*, 68.

91. Sobrino, *Christ the Liberator*, 13. On freedom, see also Sobrino, "Spirituality," 251.

of goods? In another example, circumcision of infant boys has become less common among American Christians because of parental resistance to "unnecessary" surgery. It is regarded as an issue of family choice, perhaps with medical input, but without any religious or ethical import. Yet, in Brazil, where popular culture displays anti-Jewish sentiments, Christians circumcise their boys as a visible form of solidarity with the Jews.[92]

Christian witness thus necessitates a kind of *openness to alternatives*, to challenges, to new possibilities, all of which are themselves made possible by the "new creation" in Christ. Rowan Williams writes,

> [T]o know God, it seems, involves elements of flexibility and corrigibility, not because of a trivial relativist view that what's true of God changes according to circumstances, but because of the opposite conviction, that God remains God, a "law unto himself," and, for precisely that reason, can only be discerned in the "following" of the divine action within the mutable world, in a process of learning, not a moment of transparent vision or of simple submission to a decree.[93]

Similarly, Christian witness is never complete. There are particular acts of witness, but witness is never completed by any particular action or activity. One's life is one's witness, and it is always subject to greater challenges, opportunities for creativity, and growth.[94] So, speaking of ecological ethics, Sittler writes, "in the face of unprecedented situations we need to develop creative responses that are faithful both to our inheritance from the past and to the promise of the future; we need to find new modes of symbiosis between human beings and the rest of nature that are true to the divine intentions for the unity of creation."[95] Christians cannot allow historical convention to stifle Christian praxis, anymore than they can allow the urgency of a historical event (such as the environmental "crisis") to trample the wisdom acquired through centuries of Christian witnesses.

92. Conversation with Rosalee Veloso Ewell, 2003.

93. Rowan Williams, "Hooker," 374.

94. McCabe, *What is Ethics All About?*, 97.

95. Sittler, *Evocations of Grace*, 17.

WITNESS IS COMMUNAL

Finally, Christian witness is *communal*—specifically, ecclesial. Ecclesiology will be treated in chapter 4. For now, it should be noted that Christian witness is by no means an individualistic ethic; it is not even an exclusively human ethic. That is to say, it includes more than behavior between human persons and between humans and God. The church is the primary bearer of testimony to the Kingdom of God, whether it is a tiny street-front church or a massive institutional structure. All of the aspects of the Kingdom described above—peace, abundance, reconciliation, justice, liberation—are witnessed to by the practices of the faith community. Individual Christian witnesses are welcomed, reared, and trained in that community through the sacraments, Bible study, works of mercy, and other forms of witness. As Eldín Villafañe writes, "while the church is *not* the Reign of God, yet as the community of the Spirit—where the Spirit manifests itself in a unique and particular way (Rom 8:23; 1 Cor 6:19; Eph 2:14–18)—it has the purpose both to reflect and to witness to the values of the Reign, by the power of the Spirit to the world."[96] Even the examples of prophetic individuals seeming to act on their own in dramatic fashion—Saint Francis, Martin Luther King Jr.—were nurtured and supported by strong churches. Christian witness is not about persons who happen to belong to a church; it is about churches that produce witnessing individuals.

RATIONALE FOR CHRISTIAN WITNESS

❧ After this outline of the object and character of witness, it may still be unclear why it is advantageous to understand Christian ethics as witness to the Kingdom of God. Why is it better than Christian decision-making or divine command ethics? It may not be; I am not defending witness (or discipleship) in general against other approaches to Christian ethics in general. On the other hand, this book begins with a problem: despite Christians thinking they know what to do in relation to ecological ethics, we have not managed to do it. Chapter 1 argues that part of the

96. Villafañe, "Evangelical Call," 218.

reason for the churches' inadequate response to environmental crises is theological and ethical. So I do claim that eco-discipleship is a better approach at this time to this particular problem. One justification for that claim, as explained later in this chapter, is that witnessing to the Kingdom emphasizes the sovereignty of God over all of creation and the priority of Christ in inaugurating and consummating the Kingdom. At the same time, it treats human actions as highly meaningful insofar as they establish signs to that Kingdom. Moreover, viewing Christian ethics as witness bears a strong pedigree in Christian tradition, explicates the unity of faith and action evident in Jesus and the apostles, displays the connection between creation and redemption, and, with regard to ecological discipleship, points a way between despair on the one side and messianism on the other.

To begin, we should recall that witnessing to the Kingdom of God accords with much of the Christian tradition's understanding of Jesus' own life and teachings. The phrase "witness to the Kingdom" does not occur in this precise form in the New Testament; instead, the apostles, the crowds around Jesus, and the Pauline communities are urged over and over to "follow Jesus." Following Jesus entails—as we know— obedience to God and engaging in self-sacrificial acts of love and mercy, even if they lead to suffering and death. But Jesus was not simply modeling a good life: in the life, death, and resurrection of Jesus, God inaugurated the eschaton, the new creation, the redemption of the world. This is the "Kingdom of Heaven" that was the subject of his preaching and parables. Verhey writes, "to welcome the announcement of the Kingdom of God is to allow the present impact of such a Kingdom in the ministry of Jesus to transform our values and behavior."[97] Christian witness is, therefore, responding to Jesus as the disciples were told to respond—to preach the Word, to demonstrate God's power of reconciliation and forgiveness, and to indicate the gracious abundance of the Kingdom of Heaven by their own generosity. So, "while human action does not establish the Kingdom, it is nevertheless . . . called for, and called for as the eschatologically urgent response to the action of God which is at hand and already making

97. Verhey, *The Great Reversal*, 16.

its power felt."[98] Jesus offers himself as not only the model, but also as the source of the Kingdom; in him will all things reach their consummation. Life in Christ cannot really do otherwise than testify to the radical peace, reconciliation, and abundance that is the Kingdom of God.

Moreover, witness explicates the strong unity of faith and action fitting to a world that is created, sustained, and redeemed by one God. In Gunton's words, "the heart of the matter is the relation between faith, as the gift of God, and human action as response to that gift."[99] Christoph Schwöbel elaborates,

> Christian faith and Christian action form an indissoluble unity. There is no faith that does not generate a specific form of action, otherwise we would call the presence or the authenticity of faith into question. And there is no action that does not presuppose some sort of faith, otherwise we would feel justified asking whether such behaviour could qualify as action.[100]

However, Schwöbel continues, in Christianity the faith-action unity bears a specific structure. Faith is the work of God in us, and the life of faith is our appropriating that work as the fundamental orientation for our lives. "The fundamental content of faith is what God gives to us in the threefold divine self-giving as Father, Son and Spirit in order that we should be enabled to do God's will in all our actions."[101]

Furthermore, because Christian claims are about God's power to transform earthly lives, these claims require the existence of such transformed lives. Athanasius saw this early on, saying,

> these things [in the gospel] are attested by actual experience. Anyone who likes may see the proof of glory in the virgins of Christ, and in the young men who practise chastity as part of

98. Ibid., 15.
99. Gunton, *Doctrine of Creation*, 14.
100. Schwöbel, *God, Creation, and the Christian Community*, 151.
101. Ibid., 152.

their religion, and in the assurance of immortality in so great and glad a company of martyrs.[102]

The life of the church, therefore, is the "pudding." If the church's ministry does not exemplify the possibilities of eschatological peace, then its faith is hollow. John Paul II describes faith as "a lived knowledge of Christ, a living remembrance of his commandments, and a truth to be lived out. A word . . . is not truly received until it passes into action, until it is put into practice . . . ."[103] Conversely, when the church does exemplify this "lived knowledge of Christ," when it serves as a "demonstration plot for the Kingdom of God," it evangelizes by its very presence in the world.

Christian ethics as witness also displays the christological connection between creation and redemption that chapter 2 described. After all, theology is not simply a matter of right belief. Instead, "the theologian must no longer merely defend and explain the right intellectual system but must distinguish wrong from right interpretations of how the biblical witness feeds into our contemporary social world . . . ."[104] One cannot, in fact, come to right belief, as Sobrino, Yoder, and Saint Paul note, without engaging in practice, without following Jesus with one's body and with one's life. And one cannot embrace that belief without it transforming one's life. This is what faith is *for*. In Sittler's words, "the telos of doctrine is action, the fulfillment of right teaching is not right teaching but decision and deed."[105] Christians cannot think themselves into salvation; the church cannot think itself into the Kingdom of God.[106]

So when Christians affirm the trinitarian character of the world, that all things were created and will be redeemed through Christ, they are also claiming that earthly life is in "the meantime" between resurrection and eschaton, when the Spirit is already at work sustaining and leading all things in their pilgrimage to God. That is a claim that demands a lived response, the sort of response that I am describing as Christian witness to

---

102. Athanasius *Contra Gentes* 85.

103. John Paul II. *Veritatis Splendor*, Para 88.

104. Yoder, "Biblical Roots of Liberation Theology," 61–62.

105. Sittler, *Evocations*, 47.

106. Ibid. See also Hauerwas, *Grain of the Universe*, 211.

the Kingdom. "The eschatological vision exerts ethical pressure now precisely because it holds forth a picture of a future reality which has already begun."[107] Christian ethics, especially ecological ethics, I am arguing, must be eschatological. Ecological discipleship is rooted in the past—in God's act of creation, incarnation, and resurrection—and in the future. It anticipates the New Jerusalem, where the River of Life flows through the city, the trees grow heavy with fruit, and all creatures in heaven and earth sing praises to God.

Finally, understanding Christian ethics as witnessing to the Kingdom of God provides an alternative to the messianism of "management" and the despair of powerlessness. Chapter 1 analyzes the tendency among many environmentalists (including Christian environmentalists) to describe ecological destruction in ways that entail massive, top-down managerial efforts as solutions. They frame environmental and social problems as the result of insufficient capital, outdated technology, lack of expertise, or faltering economic growth—all of which invite "managerial" solutions by international bodies (International Monetary Fund, etc.), northern corporations, northern governments, and northern non-governmental organizations—many of whom have committed the pillaging and polluting.[108] Management, too, easily turns to unjust coercion, even violence as the result of means-end analysis. On the other hand, environmentalists who distrust dominant socio-economic systems and their agents often fall into a despairing paralysis because no action by individuals or small communities can possibly make a difference. Both sides presume, usually unconsciously, that the world must be fixed or saved by humans if it is to be saved at all. Yet, that presumption constitutes an eschatology without God—a vision that the Kingdom (or ecotopia) is a human project. In contrast, if Christians affirm that God saves the world, in God's own time, they can live in hopeful witness to that reality without being tempted to wield power unjustly, and without being pulled into despair by the apparent failure of their ecological efforts. Francis Bridger writes,

107. Bridger, "Ecology and Eschatology," 295.
108. Hildyard, "Foxes in Charge," 31.

That it is God who will bring the new order and that we are merely making signposts need not act as disincentive. Rather, it frees us from the burden of ethical and technological autonomy and makes it clear that human claims to sovereignty are relative. The knowledge that it is God's world, that our efforts are not directed toward the construction of an ideal utopia but that we are, under God, building bridgeheads of the Kingdom serves to humble us and to bring us to the place of ethical obedience.[109]

The church can serve as a "demonstration plot" of ecological discipleship, making signs of eschatological abundance, peace, justice, and diversity. The church is not the Kingdom of God and can never "build" the Kingdom, but its faithful presence points to the promise of the Kingdom. The next chapter shows why the eschatological vision is necessarily ecclesial, and how the church can most faithfully testify to the Kingdom.

109. Bridger, "Ecology and Eschatology," 301.

CHAPTER 4
THE CHURCH'S ECO-DISCIPLESHIP

THE PREVIOUS chapter argues that understanding Christian life as witnessing to God's Kingdom helps alleviate the churches' dilemma between ecological despair and ecological triumphalism. Witness also emphasizes the sovereignty of God over all of creation and God's continuous activity in creating, sustaining, and redeeming that creation. Furthermore, it demonstrates the unity of faith and action inherent to the Christian tradition, and displays the intrinsic connection between creation and redemption of the world.

The Kingdom of Heaven is promised by God for the time of Christ's second coming to fulfill his lordship over all creation. As described in Scripture, the Kingdom is a realm of nonviolent peace within and among species; justice/liberation/reconciliation among all God's creatures; material abundance and sharing; righteousness; and communion with God. Christian witness to the Kingdom is undertaken in humble, yet hopeful obedience to Jesus Christ's commands: to proclaim the gospel in word and deed. This means that Christian communities do not regard themselves as executors of the divine will over creation, but as "demonstration plots for the Kingdom," positing signs to God's activity already ongoing in creation.[1] Witness is costly, requiring sacrifice and courage; it is hopeful and generative of hope; it is joyful; and it is open to alternatives and to correction.

This chapter argues that witness is a task—indeed, *the* task—for the church to undertake as the eschatological body of Christ. The centrality

1. Sauter, *What Dare We Hope*, 17. The phrase "demonstration plot for the Kingdom" is the motto for Koinonia Farm, a Christian community in Georgia dedicated to economic, social, and environmental justice.

of the church for this project is theological, rather than functional. Some theologians concerned with environmental issues express an instrumental view of the church, namely that the church's value consists in how well it contributes to environmental goals. Eschatology, however, is not a free-floating spirituality that the church happens to espouse; rather, it is the church's understanding of its role as the "demonstration plot for the Kingdom of God." The chapter then suggests particular practices of eco-discipleship appropriate to affluent churches in our time. Again, these practices cannot "save" the world; however, they facilitate and demonstrate the virtues necessary to live as witnesses to Jesus Christ in a fragmented world. They are elements of witness praxis, through which the church testifies to God's redemption of all creation. Finally, the chapter turns to the virtue of patience, which, while important for all Christian witness, is especially critical to eco-discipleship. Although patience is presented after practices in this chapter, practices and virtues are mutually dependent; we acquire the virtues through formative practices that we are better able to perform as we acquire the virtues. So we need patience to engage in ecological witness, and engaging in that witness helps us grow more patient. The church is by nature a patient institution; it can only continue its living testimony to God's sovereignty over, and providential care for, the biophysical universe through patience and perseverance.

THE MORAL AGENT OF NEW TESTAMENT ETHICS IS THE CHURCH

⁂ Recent theologians and biblical scholars note that the subject and addressee of ethical mandates in the New Testament is the faith community rather than the individual. Richard Hays carefully argues that, according to the biblical texts, "the church is a countercultural community of discipleship, and this community is the primary addressee of God's imperatives."[2] This emphasis on the church stands out most clearly in the Pauline epistles and in Luke–Acts; nonetheless, (and here I am inferring from Hays) neither the Gospels nor the Epistles present Jesus as primarily

2. Hays, *Moral Vision*, 196.

the "personal savior" of individual souls. Pointing to the faith community as the addressee of God is not a Christian innovation, of course; the Old Testament, after all, is the narrative of God's action in electing, summoning, sustaining, and saving God's *people*—the nation of Israel. Moreover, as the addressee of God's mandates and God's promises, the church is the eschatological forerunner of the New Jerusalem. The church cannot be understood, therefore, except eschatologically; it is the earthly sign of the heavenly reality. Its very existence anticipates the fulfillment of God's intentions for all of creation.

CHRISTIAN FAITH IS COMMUNAL

As Jon Sobrino points out,

> faith in Christ is essentially a community faith and not the sum of individual faiths . . . we carry one another in the faith, give our own faith and receive it, so that, formally, it is the community that believes in Christ . . . It is a feature of the act of believing that it depends on the faith of others."[3]

Because Jesus Christ is both a historical person *and* a continuing presence through the Spirit, Christians rely on the testimony of both their contemporaries and their predecessors. Even the dramatic conversions of individuals that may seem so disconnected from church life cannot occur, in fact, without the church to narrate the conversion and to support the individual's new life. This is not a limitation on God's action, for God is always active throughout creation. Rather, it is a result of the storied character of human existence. Conversion is the experience of being called to take one's part in the Christian story that is remembered, retold, and performed in the life of the church. Without the church, what would a person be converted *to*?[4]

---

3. Sobrino, *Jesus the Liberator*, 29.

4. Of course, one can point to the stories of religious leaders such as Ann Lee and Joseph Smith as contrary examples. Yet, the best account of their lives is that a new telling of an old story captured them, and the meaning of their conversions depends upon the "new" church tradition that rose up around them.

The communal character of Christianity is an anthropological claim—and one with ecological overtones. As feminists have argued for years, the liberal conception of the individual, autonomous self bears little relation to embodied, earthly existence.[5] The "I" becomes "I" only among a "we," in a communication of speech and action. Individuation does not precede association; rather it is the kinds of associations that we inhabit that define the kinds of individuals we will become. Selves, then, are socially constructed, always situated, and the context of the subject is always particular and historical. Humans are not autonomous individuals; they are embedded at the deepest level with the lives of other creatures—socially, personally, biologically. Yet, Christians take this idea a step further, due to their theological understanding of the Trinity as ultimate reality. Rowan Williams notes that everything involved in creating the social self—family, associations, environment, is created, and is, moreover, struggling to achieve its own identity against the pressures of everything else; each is what it is only in distinction from another.[6] The collection of interdependent identities is therefore a castle of cards, unable to grant the sort of security that is required for a human person's identity: the security of durability across time and place.[7] Because *everything* is created, however, the existence and identity of everything is rooted in God's freedom and generosity.

> We are here, then, we are real, because of God's "word"; our reality is not and cannot be either earned by us or eroded by others. And to say that we are unilaterally dependent on God is to recognize that God alone is beyond the precarious exchanges of creatures who need affirmation. With God alone, I am dealing

5. See for instance Benhabib, *Situating the Self.* Philosophers G. H. Mead and H. Richard Niebuhr also describe the social self. See Mead, *Mind, Self, and Society*; Niebuhr, *Responsible Self.*

6. See Rowan Williams, *Christian Theology*, 72ff.

7. Many people assume that the lack of personal identity over time, the lack of a sense of self through time, is precisely what separates non-human animals from human ones. There is considerable evidence against this, at least in the case of dolphins and gorillas. What is required for non-human identity is not, however, my point here. Rather, my focus is on the requisites for human identity.

with what does not need to construct or negotiate an identity, what is free to be itself without the process of struggle.[8]

Williams' point might seem to contradict social constructivism, indeed to support the sort of romantic pietism of a pre-existent "spiritual" self that gets called into being by God at a particular time and emerges, fully formed, at conception. Yet, God works through history, both on a cosmic scale and on a particular, personal scale. So there is not an inner, "authentic" self that stands apart from a body in time, space, and place. The self that develops through participation and negotiation with others, through the hugs and bruises of embodied life, *is* the self that is created by God.

THE CHURCH'S RELATIONSHIP TO THE KINGDOM OF GOD

❧ The church, then, is that community of worship formed by Christ as God's Son, through the intertwined movements of human events and natural history, which in turn forms its members—and sometimes nonmembers—into disciples through acts of witness enabled by the Spirit. Christians are what they are by the grace of God and the sustenance of the church, in ways both observable and intangible. The church, therefore, both local parish and church universal, is the primary addressee of Jesus' call to "follow me"—the call to witness to the Kingdom of God. This means that Christian language about the role of the church is subject to the same constraints that were outlined in the last chapter. The church points to the Kingdom and posits signs of the Kingdom, but does not bring about or usher in the Kingdom. This claim correlates with the earlier argument about human action in general. However, part of the Christian tradition has held that the church is somehow instrumental in the establishment of the eschaton. The letter to the Ephesians, for instance, stresses the cosmic significance of the church, "which is described as the fulfillment of God's design to gather all things together in Christ. God has 'put all things under his feet and has made him the head over all things *for the church*, which is his body, the fullness of him who fills all

---

8. Rowan Williams, *Christian Theology*, 72.

in all'" (Eph 1:22–23).[9] God will complete the reconciliation of the universe, on Richard Hays' interpretation of Ephesians, *through* the church. This understanding of the church's role may make it essential to God's purposes in a way I want to avoid, not because I dispute the centrality of the church—that is part of my overall argument—but because, given the predominant theological worldview, overstating the centrality of the church is as much a risk as understating it. First of all, as Northcott writes, the "transforming work of the Holy Spirit is not limited to the bodies, households and churches of Christians. The Spirit is said to be already at work in the creation, drawing human history and created order to its destiny of final fulfillment in the eternal plan of God which is revealed in Jesus Christ."[10] So Christians cannot stress the role of the church in redemption to the point that either the redemption of creation or the radical materiality of human life is lost (again). As Schwöbel writes, "the Christian community has the same ontological constitution as creation as a whole."[11] That is, the church, like creation as a whole, "owes its being to the absolute giving of the Triune God . . . being made righteous is, just like being created, an absolute gift." And both the created universe and the church have their end not in themselves, but in the Kingdom of God.[12] The church is distinct from the world because it is the body of Christ, but the whole created order is under Christ's governance. So God indeed works through the church, but not only through the church.

I am arguing against the notion that the church is a vehicle to carry people toward responsible eco-behavior as defined non-theologically (Oelschlaeger), but also against the Eastern fathers' idea that all of creation is saved *through* human worship. The church neither establishes the Kingdom of God by its activities of ecological discipleship, nor extends the efficacy of Christ's saving power beyond the human species, because Christ alone has already saved the world. The point for Christian ethics is that the church's mission is to witness to the God whose lordship extends

---

9. Hays, *Moral Vision*, 62. His emphasis.

10. Northcott, *Environment and Christian Ethics*, 202.

11. Schwöbel, "God, Creation, and the Christian Community," 170.

12. Ibid., 171.

far beyond the church. The task for the church is to be in the world as a living image of submission to that lordship.[13]

The church, therefore, is not just another social-action organization, or another neighborly community in the existing political structure, but an alternative polity, "the civic assembly of the eschatological city."[14] As an eschatological presence in the world, it presents a contrasting model of both history and politics. "Civil" communities presume a secular model of history in which events occur in the past, the present, or the future, and continue to an ever-retreating horizon (or perhaps an apocalyptic cataclysm, in some anti-modernist scenarios). In modern time the present tends to kick off the traces of the past and lunge toward the future; in postmodern time both present and future are fragmented and scrambled, but any connection with the past is still only momentary. Moreover, the effects of globalization encourage people to imagine/construct connections across vast geographical distances, yet time is continually compressed until what counts as "the present" becomes of smaller and smaller duration. In contrast, Christian time is finite—not only the events that occur in time, but time itself—to be ended at the hour of God's choosing. And in the Eucharist, past, present and future are brought together through Christ, as if he were gathering the folds of time in his hand. Cavanaugh writes,

> [I]t is in the eucharist that Christ himself, the eternal consumma-tion of history, becomes present in time. The eucharist is not a mere reoffering of Christ's sacrifice by the priest before the watch-ful eyes of the faithful. The earthly eucharist is the eternal action in time of Jesus Christ himself, "high priest, one who is seated at the right hand of the throne of the Majesty in the heavens." (Heb 8:1)[15]

Christian time is commonly perceived as linear; linear time is indifferent, mechanical, and straightforward. Christian time, however, is eschatologi-cal: a gift with a purpose, a "space" in which God's intentions are made known. It is not, though, a route to the Kingdom such that each year

13. Cone, *Black Theology of Liberation*, 132.

14. Yeago, "Messiah's People," 150.

15. Cavanaugh, *Torture and Eucharist*, 223.

is closer than the year before; the relationship between the present and future is not calculable, but a matter for prayer, because earthly time itself is created by God.

THE CHURCH AS EUCHARISTIC COMMUNITY

❧ Moreover, the Christian community does not form itself, it is formed by Christ through the Eucharist. Richard Hooker understood that in Eucharist Christians are conformed to Jesus Christ through participation in his sacrifice; and that conformity—the taking up of all members into Christ's body—enacts the church as the body of Christ.[16] Harvey Legrand points to the first reference to Eucharist in 1 Corinthians 16–17: "The cup of blessing that we bless, is it not a sharing in the blood of Christ? The bread that we break, is it not sharing in the body of Christ? Because there is one bread, we who are many are one body, for we all partake of the one bread." In this passage the eucharistic body precedes and causes the ecclesial body.[17] Rather than describing the church as a community that performs Christian rituals, then, we may say that the performance of Eucharist creates and re-creates the community.

Because the church is repeatedly enacted through Christ's sacrifice on behalf of the world, the event in which all things are forgiven and made new, the church is predicated on the possibility of reconciliation and forgiveness. Again in contrast with a functional description, the church at its best is not a community that forgives, but forgiveness communally embodied. As George Hunsinger writes about the confessing church,

> [It] rejects the individualism of the conversionists, the secularism
> of the activists, and a matter peculiar to them both: the equating
> of faithfulness with effectiveness . . . it is committed first of all to

16. Rowan Williams, "Hooker," 322. "[I]n the Lawes, ritual acts don't "declare" or signify gemeinschaft but create it. Compassion, fellowship, and mutual love seem possible only when relations among persons are mediated by a "ghostly fellowship with God and Christ and Saints" (Hooker, *Lawes of Ecclesiastical Politie*, VIII.4.6 quoting 1 John 1:3).

17. Legrand cited in Kärkkäinen, *Introduction to Ecclesiology*, 31. The idea that the "Eucharist makes the church" is common among early church fathers and contemporary Orthodox theologians.

restructuring neither society nor the heart but the church according to the will of God. The church restructured will be a church
of reconciliation, a church of nonconformity, and a church of
the cross.[18]

In this passage Hunsinger points to reconciliation as essential to the
church, just as it is to the Kingdom of God. For the church to act as a
sign of the Kingdom, it must first live into the gift of forgiveness and
reconciliation granted in Christ. This means that the liturgical confession
of sin is not primarily an individual confession, but a corporate one:
the faith community has sinned. Yet, God's forgiveness also means that
the church is not bound by its sinfulness, but is enabled through the
Spirit to live into new possibilities of peace and abundance. Therefore,
the hegemony of gratuitous violence to non-human creation cannot
prevent the church from exploring nonviolent—or at the very least less
violent—relations with its earthly community. Carol Adams writes of
the difficulty of interrogating meat-eating practices in a meat-eating
culture; the discourse itself is suppressed by the overwhelming presumption that meat-eating is normative.[19] Challenging the dominant political
ethos is nearly impossible from a position within the dominant political
structure. Likewise, introducing practices that foster peaceful relations
in contemporary Western culture is extremely difficult. The church,
however, stands outside secular politics. The incarnation, ministry and
resurrection of Jesus Christ makes peace more fundamental than war;
and it makes love more fundamental than hatred, so there is always the
Spirit-led possibility of interrupting the culture of violence with pockets
of reconciliation.

The Eucharist provides both the time and the demand for such in-
terruptions, as the Kingdom itself irrupts into the present. As Douglas
Farrow points out, "the eucharist lends to the church its eschatological
dynamic, as a participation both in the brokenness of the crucified and

18. Hunsinger quoted in Hütter, "The Church," 49.

19. Adams, *Sexual Politics of Meat*, 91. Adams also makes the important point that
Western humans' most common experience of nonhumans is in eating them.

in the victory of his resurrection and ascension to the Father."[20] The Eucharist is both performance and gift: re-enacting the story that forms the world and the church, and a "foretaste" of the eschatological banquet. Eucharist forms, reforms, and sustains the faith community, and thus makes an insistent sign in the world of the intruding presence of God's Kingdom.

THE CHURCH IS NOT AN ACTIVIST AGENCY

❧ Christian witness, therefore, entails both "works" (outreach, ethics, and evangelism) and worship. They are not activities that the church does; rather, they are performances and gifts from God that constitute the church as the eschatological body of Christ.[21] The two aspects of church—worship and works—are not identical, but are deeply interconnected, for neither can be true without the other. Worship "is the most regular way that most Christians remind themselves and others and are reminded that they are Christians. It is the most significant way in which Christianity takes flesh, evolving from a set of ideas and convictions to a set of practices and a way of life."[22] A church that becomes distinct from the world only by its formal liturgies and the shape of its building cannot testify to the Kingdom, because it never speaks beyond its own walls—and if it is not proclaiming to the world, then it cannot worship rightly. Conversely, a church that loses itself in "outreach" ministry risks confusing its identity with that of a secular activist agency. So Cone is correct when he says,

> the church can neither retreat from the world nor embrace it. Retreating is a complete misunderstanding of the Christ-event, which demands radical, worldly involvement on behalf of the oppressed. Retreating is . . . a luxury that oppressed people cannot afford . . . Embracing the world is also a denial of the gospel . .

20. Farrow, "Eucharist, Eschatology and Ethics," 201.

21. Martin Luther seems quite correct on this point, although he employs his characteristic exaggeration, that both Eucharist and good works are God's gifts to us rather than our offerings to God. See his "Freedom of a Christian," 42–85.

22. Hauerwas and Wells, "Christian Ethics as Informed Prayer," 7.

. identifying the rise of nationalism with Christianity, capitalism with the gospel, or exploration of outer space with the advancement of the Kingdom of God serves only to enhance the oppression of the weak.[23]

Unfortunately, Cone goes on to say that "Christ's church is to be found where wounds are being healed and chains are being struck off. It does not matter whether the community of liberators designates their work as Christ's own work. What is important is that it is the human thing to do."[24] Here the church disappears into, or is reduced to, its justice mission. This move enters treacherous territory, for without the church there is no way to determine whether the "justice" in question is God's justice or some human (mis)understanding of justice. Moreover, the disciples are commanded to work *and* preach, to demonstrate the Kingdom and to identify it as Christ's work.[25] Jesus himself not only liberated and healed, but he also prayed and worshipped. The church is not just another social agency or another activist collective; rather, it is the body of Christ in history, and must reaffirm this identity at every step. Work of liberation where "wounds are being healed and chains are being struck off" may well be the work of the Holy Spirit, in accordance with God's intentions, but such work is *not* the church. This liberative work should not be shunned or condemned by Christians. It may provide real relief of suffering and hope for a better future, as well as providing useful examples to consider or possible partners for church efforts. When the church, however, undertakes a "new" form of ministry, such as environmental protection and healing, it must not understand its mission as sending out masses of congregant members to join secular movements, or "inspire" them toward greater good.

23. Cone, *Black Theology*, 132.

24. Ibid., 134.

25. Matt 10:5–8: These twelve Jesus sent out with the following instructions: "Do not go among the Gentiles or enter any town of the Samaritans. Go rather to the lost sheep of Israel. As you go, preach this message: 'The Kingdom of heaven is near.' Heal the sick, raise the dead, cleanse those who have leprosy, drive out demons. Freely you have received, freely give."

An especially egregious example of this instrumentalist view of Christianity is an *E Magazine* editorial, "Save the Earth, Not Just Souls," (Nov/Dec 2002). The editor claims that church members may feed and clothe the poor, but rarely engage in any discussion:

> Critical of the forces that cause and perpetuate such suffering and poverty in the first place. For this reason I can't help but see organized religion as a major accomplice—through the sin of silence—in the continued economic hopelessness of a growing majority of people on this planet, a hopelessness that also causes environmental degradation."[26]

He then applauds the work of "pioneers" who initiate dialogue among religious leaders on environmental issues. However, "the major religions will need more than gentle persuasion to enlist them in efforts to rescue the environment—or to be a force for any kind of real progressive change." And he concludes, "it's time the churches stop contemplating their navels and plying only the narrow focus of personal salvation. It's time they got on with the business of saving the world they claim God put them in charge of." This editorial expresses a purely functionalist view of religion: a religion's role is to effect "progressive change" in its cultural context. The writer cannot envision the possibility that the church enacts a different culture entirely. To be fair, the churches have often failed in living out this alternative Christian reality. So, for an atheist such as the editor, the only conceivable "function" for the church is that of cultural critic—a function it cannot or will not fulfill, in his limited experience.

Even though most environmentalists are more respectful of religion (or at least of some religions), they often share *E Magazine's* functionalist perspective. Even more problematic is that some Christian environmentalists themselves adopt this viewpoint. Portions of Larry Rasmussen's work in *Earth Community, Earth Ethics* leans in this direction, as indicated in chapter 1.[27] His older book, *Moral Fragments and Moral Community*,

---

26. Editorial, *E Magazine*, Nov/Dec 2002.

27. Charles Pinches remarks about Rasmussen's book, "He is principally concerned not with whether an idea or text is theologically astute but whether it moves our spirits" in eco-friendly directions. Pinches, "Eco-Minded: Faith and Action," 756.

is a thoughtful and thought-provoking analysis of the disintegration of modern/postmodern civil society and the sorts of communities that are required to re-construct it—not in premodern form, but in ways that embrace the religious, political, ethnic, and racial diversity of a globalized world. He provides an excellent description of the church's role in moral formation of Christian disciples. Yet, here too, the success or failure of the church fulfilling that role is regarded more for its assistance in the "larger" project of enabling civil society to function than for its faithfulness to God's commands. This point should not be overstated, for Rasmussen is not a reactionary who wants the church to produce docile citizens of secular capitalist regimes. The vision of civil society he presents is one that includes many virtues endorsed in the Christian tradition. It may not be, however, the only "good" society for Christians. Moreover, Rasmussen treats the church purely as a "moral community," and thus formally the same as any other voluntary community with rich tradition and ritual practices. Rasmussen's understanding of the church is, at bottom, sociological rather than theological: a moral community rather than the body of Christ.[28] The church is both, of course, but its primary identification must be understood and enacted as the body of Christ for its role as "moral community" to be carried out with any faithfulness.

A church that loses itself in social ministry also risks confusing its signs of the Kingdom with the Kingdom itself, or with the onset of the Kingdom. This was the flaw of Rauschenbusch's "Social Gospel," evident when he writes,

> The saving of the lost, the teaching of the young, the pastoral care of the poor and frail, the quickening of starved intellects, the study of the Bible, church union, political reform, the reorganization of the industrial system, international peace—it was all covered by the one aim of the Reign of God on earth.[29]

This confusion of the church's witness with the object of its witness follows directly from the failure to acknowledge God's ultimate freedom as

---

28. See Rasmussen, *Moral Fragments*, chapters 7 and 8.

29. Rauschenbusch, *Christianizing the Social Order*, 93.

creator. Were the church to disappear suddenly, God would still redeem the world. The church serves the Kingdom through its witness; it does not deliver the Kingdom into the world. Contrast this with James Nash's argument:

> [The church's] ecclesial responsibilities, however, are more than *anticipations* of the divine Reign. Here the social gospel in North America offers and important corrective to some current eschatological emphases . . . The church exists for the sake of the Kingdom, but its task is more than an anticipation of the Kingdom; it is the actual but provisional construction or creation of the Kingdom on earth.[30]

Nash's justification for this claim is that "ethical and ecclesiastical achievements must be more than anticipations if they are to have eternal significance or meaning . . . if our historical existence and moral acts of liberation and reconciliation are to have enduring value, they must in some sense be contributions to, preparations for, and participations in God's final re-creation."[31] What Nash (and Rauschenbusch) misses is what we may call the "creaturely paradox." That is, there is *no* meaning or value in earthly life other than that given by God's grace. Yet, because God through Christ is the author of all, everything already has eternal significance, by participation and reception of God's creative activity. Making signs of the Kingdom (or, to use Nash's term, "anticipations") is not mere busy-work, or some superficial imitation of the "real" Kingdom-building. According to Christian Scripture and tradition, *all* goodness of earth will be preserved (in some way) in the eschaton—otherwise, redemption could not be complete. So Christians need not fear that their witness somehow lacks significance in the divine plan—everything is significant. Christians, however, are not witnessing to ourselves, which is one of the dangerous implications of Nash's argument. The point is not whether human actions have ultimate meaning, but that God is the source and substance of meaning. Christians do not do God's work in the sense that

30. Nash, *Loving Nature*, 136.
31. Ibid.

we are substituting for God; rather, Christians do the work that God has given us to do.[32]

The work we are given is essentially an extension of our worship: by its witness, the church participates in the chorus of praise offered by the whole creation.[33] This understanding of Christian life as worship bears particular implications for environmental concerns. Bauckham writes, "the creation worships God just by being itself, as God made it, existing for God's glory. Only humans desist from worshipping God; other creatures, without having to think about it, do so all the time."[34] This is only

---

32. The Post-Communion Prayer reads:

> Almighty and everliving God,
> we thank you for feeding us with the spiritual food
> of the most precious Body and Blood
> of your Son our Savior Jesus Christ;
> and for assuring us in these holy mysteries
> that we are living members of the Body of your Son,
> and heirs of your eternal Kingdom.
> And now, Father, send us out
> to do the work you have given us to do,
> to love and serve you
> as faithful witnesses of Christ our Lord.
> To him, to you, and to the Holy Spirit,
> be honor and glory, now and for ever. Amen.

Episcopal Church, *Book of Common Prayer*, 366. Nash makes no mention of Eucharist in his book, and scarcely mentions sacraments except for the sacramentality of nature. This is a critical omission. It is interesting that Rasmussen, Oelschlaeger, and Nash all give minimal (if any) attention to Eucharist, *and* all three get the relationship between ecclesiology and eschatology a bit skewed. This fact alone supports my emphasis on the crucial role of Eucharist in the church's life and mission. Rasmussen, *Earth Community*; Oelschlaeger, *Postmodern Environmental Ethics*; Nash, *Loving Nature*.

33. Saint Basil prayed, "O God, enlarge within us the sense of fellowship with all living things, our brothers the animals, to whom you gave the earth as their home in common with us. We remember with shame that in the past we have exercised the high dominion of man with ruthless cruelty, so that the voice of the earth, which should have gone up to you in song, has been a groan of travail. May we realize that they live not for us alone but for themselves and for you and that they love the sweetness of life." Quoted in Wynne-Tyson, *Extended Circle*, 9.

34. Bauckham, *God and Crisis of Freedom*, 177.

partly right, for we cannot ignore the fallen character of creation—the cruelty of animals may not be deliberate in the same way human cruelty is, yet it is similarly offensive to God's redemptive purposes.[35] Bauckham, however, is right to say,

> there is no indication in the Bible of the notion that the other creatures need us to voice their praise for them. This idea, that we are called to act as priests to nature, mediating, as it were, between nature and God . . . intrudes our inveterate sense of superiority exactly where the Bible will not allow it.[36]

If the human-as-priest motif is understood to mean that the human provides a necessary mediation between non-human creation and God, it violates the biblical understanding of God's intimate relation with all of creation. This is the difficulty with, for instance, Zizioulas' appropriation of the Greek Fathers' idea of humans as the microcosm of the macrocosm.[37] On the other hand, if human-as-priest is taken as a christological mandate for a life of service to *all* of creation, as in Andrew Linzey's work, it enables a witness to Jesus' self-sacrifice for the sake of the whole world.[38]

ECOLOGICAL WITNESS

Thus far, this chapter has argued that the addressee of Christian ethical mandates is the church, and that the church is that peculiar community that anticipates through worship and works the consummation of God's reign. All the church's "business," therefore—its worship, works, institutional structures, finances, and so forth—must be formed eschatologically.

---

35. Some animals are clearly capable of conscious cruelty, just as they are of deliberate kindness. Ironically, these are not the animals usually brought for "blessing" on Saint Francis' Day, because they do not make good "pets."

36. Bauckham, *God*, 228.

37. Zizioulas, "Preserving God's Creation: Three Lectures on Theology and Ecology." Thomas FitzGerald also takes this line, though Zizioulas is more cautious and precise in his reasoning. FitzGerald, "The Holy Trinity and Creation."

38. Linzey, *Animal Theology*, 32.

In other words, the church is not an institution or community that happens to hold eschatological beliefs; rather, it is created and sustained eschatologically. It is pulled into existence in the present by the future eschaton.[39] Everything it practices, including its relations with the earth, should witness to Christ's coming redemption of the world. This means, first, that witness is an inclusive category for the church, not limited to "outreach" or "evangelism," and second, that churches must always be mindful of the characteristics of the Kingdom. Especially in a "crisis," such as the "environmental crisis," it is easy for churches to become diverted from their mission onto a problem-management path.[40] Yet, as shown in earlier chapters, that path frequently leads to an implicit rejection of God's sovereignty and a hopeless reluctance to undertake difficult, long-term work.

The church, then, is called to witness to Christ's inauguration of the Kingdom of God, and to that Kingdom as the consummate realm of peace, abundance, justice/liberation/reconciliation, and righteousness for all of creation. The previous chapter delineated how those aspects of the Kingdom pertain to non-human creation; the next section of this chapter indicates ways that churches can witness to them. These examples are not exhaustive, but they are meant to engage the imagination. On the other hand, these examples should be taken seriously, for eco-discipleship practices should accord with God's redemption of creation as narrated in chapters 2 and 3.[41]

This section does not read like "academic theology," with its discussion of fertilizer and bread. Nonetheless, such mundane matters need to be seriously—and theologically—addressed for congregations to engage in the sort of eco-discipleship I advocate. It should be noted here that these practices are suggested for affluent churches whose members do not

---

39. Jürgen Moltmann also uses this kind of language. For Moltmann, however, history sometimes seems a category prior to God; God is that which demonstrates how humans should transform history, and enables them to do so. See Sauter, *What Dare We Hope*, 132ff.; Moltmann, *Theology of Hope*.

40. See the section on patience, below.

41. I guard against the inclination to select some convenient "green" option rather than engage in real, disciplined change.

struggle to meet basic needs of food, work, or shelter. Churches in more straitened circumstances may find their witness needs to take a different direction. Feeding hungry parishioners, for instance, may take priority over choosing particular foods to offer.

What emerges in the next section is that Christian witness to Christ's redemption of creation differs from secular environmental activism, even though the two overlap in several areas. Christian eco-discipleship is neither primarily about the churches' urging members to join environmental organizations, nor about enacting regulations to protect "natural resources" for human use. Rather, Christian witness to the Kingdom entails incorporating practices in church life that demonstrate the possibilities of Christ's redeeming work. Christians cannot "save the earth," but they can exhibit, in small ways, what a "saved" earth might look like.[42]

PEACE, ABUNDANCE, AND JUSTICE

Let us consider eschatological peace, abundance, and justice as they affect non-human creation. Peace is, perhaps, the most radical characteristic of witness. Much of Western culture—in particular, Western science—is predicated upon the assumption that humans and "nature" are at war, a war that is continuous and bloody and that requires extreme methods for humans (usually men) to win.[43] After Bacon and Descartes, natural phenomena were studied less as clues to the divine and more for ways they could be conquered or "turned" to human use. Living in peaceful witness, therefore, is not equivalent to living "nicely," refraining from causing offense, especially because so much of human-nonhuman violence has been masked by historical convention. Instead, eco-discipleship

---

42. Obviously, this comment, and the whole book, plays off the different meanings of "saved." For secular environmentalists, "saving" means preserving the earth from despoliation, for the future use of its inhabitants (and this is a laudable goal). For Christians, however, "saving" means drawing into the eternal life of the Triune God, which has already been accomplished—though not completed—by Jesus Christ.

43. Merchant, *Death of Nature*, 164ff.; Harding, *Science Question*, 113ff.

requires radical shifts in action that might appear alarming or suspicious to the church's human neighbors or even to some church members.

Peaceful relations with non-human creation requires, at least, a minimizing of violence toward the parish's local land, creatures, and eco-systems. The intensive use of pesticides and herbicides causes undeniable long-term damage to ecosystems and the health of human agricultural workers. And the eradication of forests leads to global warming, hotter micro-climates, unbalanced local ecosystems, and threats to plant and animal species. Therefore, rather than aiming for the "green" lawns of English country estates (themselves indicators of wealth and earthly domi-nance), churches should aim for "green" practices of fostering native plant habitats, organic fertilizers (if required), wildlife feeding stations with ap-propriate, minimal use of non-native plant species, and the most gentle "development" of church property—mulch paths rather than sidewalks; gravel screen rather than paved parking; trees and underbrush uncut. In this way the church can physically present a (partial and imperfect) sign of the Kingdom, where species live as neighbors rather than enemies.

Peaceful witness also requires the minimization of animal death at the hands of the church.[44] Rather than church barbecues or pig pickings, vegetarian meals should be promoted and served whenever possible. If meat is served, it should be obtained from local organic sources instead of distant factory farms, so the church is not supporting food production

---

44. For an excellent historical analysis of meat-eating and vegetarianism, see Carol Adams' *Sexual Politics of Meat*. For good Christian arguments on vegetarianism, see also Andrew Linzey, *Animal Theology*. Richard Bauckham argues that "an argument that meat eating *is* absolutely wrong would clearly contradict the Christian belief in the sinlessness of Jesus" and "would also cut Christianity's roots in the Jewish tradition of faith to which Jesus so clearly belonged" (Bauckham, "Jesus and the Animals II," 54). In fact, most Christians are not in a position to consider an absolute prohibition against meat-eating, because they have not taken the first steps toward reducing their meat-eating. It is as though someone just embarking on a spiritual pilgrimage tries to leap the final obstacle, before acquiring any spiritual fitness or experience. Moreover, questioning the ultimate step of a particular discipline can serve to reject any criticism of current practices. Most Christians need to hear the basic mandate of nonviolence toward God's creatures well before they can consider vegetarianism as a matter of faith. See also the comments on pedagogy in Yoder, 'Patience' as Method," 25–26.

systems that create widespread animal misery and suffering.[45] The liturgical traditions of fast days can be recovered as a communal parish practice. Occasional fasting not only prepares the body for prayer, but testifies to the eschatological abundance that enables Christians to refrain from hoarding out of a fear of scarcity. Concomitantly, witnessing to peace and justice precludes participating in food production systems that harm human workers. It is nearly impossible for an American church or home to extricate itself from food production systems that rely on low-wage labor, poor working conditions, environmental degradation or international inequities. Yet, more alternatives are becoming available. "Fair trade" shade-grown organic coffee, for instance, that is priced higher to provide real remuneration to the growers in Latin America and Africa, is now commonly available and should be used in every parish.[46]

The bread and wine offered at Eucharist are both the broken body and the blood of Jesus, offered in nonviolent obedience to God, and the offering Christians make back to the Lord in thanks. Regardless of a church's particular eucharistic theology, the elements of bread and wine are the most significant material expression of Christian worship. It is crucial, therefore, that what we offer to God and take into our bodies as nourishment be a testament to God's providential care for the earth, and Christ's redemption of all creation. Yet, most eucharistic wine has been produced at the expense of the health and well-being of vineyard workers and the land.[47] While the plight of migrant workers has improved in California, the largest North American source of wine, it still commonly

45. Vegetarianism is not the focus of my work here. Carol Adams, however, makes an excellent point about modern methods of meat production. She writes, "we come to see that a piece of "meat" turns the miracles of the loaves and fishes on its head. Where Jesus multiplied food to feed the hungry, our current food-producing system reduces food sources and damages the environment at the same time, producing plant food to feed terminal animals . . . " (Adams, *Sexual Politics*, 156). In effect, factory farming creates scarcity where abundance formerly occurred.

46. For information on the philosophy and practices behind fair-trade coffee, see Kearns, "What Does Justice Taste Like."

47. Boucher-Colbert makes a similar point, but he couches it in terms of felt symbolism, as if the Eucharist is only meaningful if we find it so. Boucher-Colbert, "Eating the Body," 127.

includes lack of sanitation, adequate housing, protection from chemicals, or anything approaching a just wage.[48] This issue is a difficult one for churches at the parish level. Unlike, say, the auto industry, which is scrutinized by a number of outside parties who monitor its products and claims, the wine industry is nearly opaque. A single wine company, such as Gallo, gathers grapes from its own vast vineyards as well as other wine growers. Then, to reach different markets, it sells its wine under a variety of different labels, many of which do not mention the Gallo name. As a result, it is nearly impossible to trace a bottle of wine to a particular vineyard or farm, or to identify the practices by which it was produced. On the other hand, the market for organic wine is emerging, and more growers are adopting organic growing practices.[49] Within the next decade, it should be possible for churches to obtain port wine grown from organic grapes at affordable prices.[50] At present, churches can at least try to get the "least bad" wine. For instance, Napa County has enacted some of the strongest requirements for growers to provide housing for farm workers, so from a justice perspective, Napa wine is better than Sonoma wine. Some wineries use "integrated pest management," which focuses on using lesser amounts, and the least harmful, chemicals.[51] And smaller wineries tend to treat workers more fairly than large corporations. Even if churches cannot locate organic, fair-trade wine, the act of asking the questions may eventually encourage wineries to improve their production methods. This kind of work, inefficient, tedious, and without satisfying

48. Sanchez, "In California's Vineyards, Grapes of Wealth and Wrath," A03.

49. In the U.S., "USDA organic wine" must not have any sulfites added. "Made from USDA organic grapes" means that sulfites have been added to increase chemical stability of the wine. The Archbishop of Canterbury, Rowan Williams, just issued a call to Anglican churches in England to use organic wine at Eucharist. Perhaps a larger market will increase the availability of organic wines. Rowan Williams, "Sharing God's Planet."

50. Port is generally used for communion wine, at least in Anglican churches. It has a high alcohol content (fifteen percent), which kills germs in the chalice, and the red color matches the color of blood. I am told that Roman Catholic parishes are more flexible in their wine requirements. If so, this is an interesting fact, though not relevant to the argument here—does that flexibility contradict or stem from a different eucharistic theology?

51. Glionna, "Napa Growers," B1; "Migration Dialogue," website at University of California-Davis: http://www.migration.ucdavis.edu. (Accessed March 21, 2005.)

results, is the work of witnessing to the peace and justice of the Kingdom of God—the work of discipleship.[52]

Communion bread or wafers, too, should serve as a testament not only to Christ's sacrificial death, but to Christ's uniting all things under his reign such that hardship, oppression, and exhaustion of the land no longer factor into food production. Bread presents less of a dilemma than wine in terms of fair labor practices. Wheat production has become so automated that little or no manual labor is required. (This itself is probably not an improvement, in theological terms.)[53] So the concern for Christians is not so much about labor as about land. Conventional wheat production exhausts soil fertility, contributes to erosion, and requires increasing application of chemical herbicides, pesticides, and fertilizers. Organic bread, however, is readily available for communion, or can be baked by volunteer parishioners. Home-baked organic bread concords with Eucharist theology, in which the work of human hands on the fruits of the earth make up the sacrifice we offer to God.[54] Again, the point is that our offerings to God ought not to be the toxic fruit of harmful agricultural methods, but products of careful, attentive working of the land. This should not be taken as a yearning for moral purity, as if there were some way of sanitizing Eucharist beyond all human sin, or as if the "right" offerings secure a higher level of faithfulness. Rather, for the church to be a living sign of the Kingdom of God, it must continually strive to demonstrate the possibilities of justice, peace, and abundance. This means, in this case, the church can—and should—point to the es-

52. In 2005, the least expensive organic port, and the only U.S. organic port, is from Badger Mountain in the Pacific Northwest. It sells for about 3–4 times the cost of an inexpensive conventional port. Still, the increased cost is a minor item in many churches' budgets.

53. See Wendell Berry and Wirzba, *Art of the Commonplace.*

54. As of 2007, Holy Trinity Altar Bread makes organic communion wafers. "Each batch of bread is made prayerfully by workers who are dedicated to serving God and providing His Church with the finest quality altar breads. Our staff is committed to making bread in the canonically prescribed tradition, using only pure water and organic wheat flour." (Holy Trinity, http://holytrinityaltarbread.org/about.html (accessed October 18, 2007). They are located at 3664 North Delaware Street, Indianapolis, Indiana 46205. Phone: (765) 475-4970. Other sources of organic altar bread are becoming available as well.

chatological possibilities of sharing the bounty of earth without destroying the land or profiting at the expense of its human laborers.

The abundance of the Kingdom is far beyond human imagining, so much so that in our day, at least, humans strive to collect more and more possessions as security against the dreaded prospect of scarcity. Living in the grip of individual and structural sinfulness, Christians are not immune to this behavior, but it fails to demonstrate a faithful confidence in the promises of God, or in the fundamental fellowship of God's creatures. Living into the Kingdom's eschatological abundance requires, against our intuitions, a kind of joyful austerity. Yet, Christians also know (less so now than in the early centuries, perhaps) that following Christ entails openness to suffering—not for its own sake, but to follow Christ's obedience to God. Christians are freed, therefore, to undertake difficult and apparently inefficient labors of obedience, knowing that some discomfort may strengthen their virtue and offer a model for others. Conserving the energy used for climate control, and using sustainable energy sources where possible, are practices churches can undertake in this direction.

Clearly, these steps toward peace and justice are only the beginning. Yet, already the church's "green discipleship" looks quite different from much secular environmentalism. The church's witness is not so much about enacting regulations that protect natural resources as it is about demonstrating an alternative way of life in Christ. As shown in chapter 4, this witness is worshipful—joyfully confessing the sovereignty of God. It is also guided by the dictates of faithfulness rather than efficacy, hopeful and hope-generating, likely to be costly, and suffused with the memory of previous Christian witnesses. It needs to be courageous, creative, self-questioning, patient, and perseverant. Thus are we brought back to the Christian virtues, the gift and development of which enable the church to live into its witness.

PATIENCE: THE ECOLOGICAL VIRTUE

❧ In addition to faith, the central virtue for ecological discipleship is patience. The prima facie evidence for this claim is the obvious role that impatience has played in environmental destruction: the impatient ambition for absolute dominance over the "natural" world, the impatient disregard for biophysical limits, and the impatient greed for profit.[55] Patience is by no means a cure-all for human destruction of God's creation; however, it is a necessary virtue for the development of a vigorous witness to God's care for the world.[56] Ecological witness or Christian activism cannot come about in the absence of patience. Even the word "witness," in one of its meanings, signifies watching—and watching is an activity of patience.

Given the central role played by patience in eco-discipleship, then, it is important to understand the nature of patience, the ways in which it differs from passivity, and how it functions as a peculiarly Christian virtue with regard to ecological witness. David Baily Harned argues in favor of re-establishing patience in its central position among Christian virtues, where it can stand as an urgently needed corrective to the extreme *impatience* of the modern world.[57] He draws on numerous resources in the Christian tradition: Tertullian, Augustine, Thomas á Kempis, Aquinas, and Calvin all emphasize the role patience plays in Christian lives. Harned argues for the benefits of patience, the evils of impatience, and the interconnectedness of patience with other theological and moral virtues: faith, hope, love, prudence, courage, justice, temperance, gratitude, and humility. While Harned overstates his case—a suspiciously large assortment of Christian qualities is crowded under the "patience umbrella"—he makes a good historical assessment of patience in the Christian tradition.[58]

Unfortunately, Harned anchors patience in human distinctiveness, rather than on the nature of creatureliness. He writes that waiting is an inevitable part of life, rather than a sign of human failure. "What is most important, however, is the recognition that waiting is at the center of

55. Agrarians Wes Jackson and Wendell Berry often speak about virtues such as patience.

56. For a careful analysis of the virtues of temperance and prudence in relation to creation, see Davis, "Preserving Virtues."

57. Harned, *Patience: How We Wait Upon the World.*

58. For instance, Harned writes that patience is the only judge of when impatience is the appropriate response to suffering or injustice. Yet, surely that role belongs to prudence, formed by courage and charity, (Ibid., 150).

things not because of the nature of the world, not because of the nature of society, but because of who we are—made to be dependent, incomplete in isolation from others."[59] That is, humans are created in the image of a God who is Creator, Redeemer, and Sanctifier in eternal communion, so interdependence and love are essential to the very being of the divine. And such interdependence makes patience necessary because each person must, in some way, rely upon the presence or activity of another. In contrast with Harned's anthropocentrism, however, every part of creation reflects the creator in exactly this way: from the atom to the solar system, each created entity is interdependent with its neighbors. Sociality and dependence are not distinctive to human beings. In fact, regarding interdependence as a human distinction undercuts the very idea Harned is trying to articulate, for the interdependence he emphasizes is not limited to humans, but unites creatures of widely differing species.[60] The effect of Harned's misstep is, ironically, to isolate humans from their non-human neighbors and biophysical environment.

This anthropocentric isolation is compounded when Harned writes, "faith in the creator demands unwearying and persevering effort in the confidence that the universe we have been given is a benign and appropriate context for the exercise of human liberty."[61] Taken at face value, the statement is true, but it strongly implies that liberty is the most important aspect of being human, and that the universe was designed for the sake of that liberty—both claims that contradict the Christian doctrines of creation and sin.

In contrast, patience *is* woven into the web of the universe because this universe reflects—in dim and fractured ways—the divine patience of its creator. Part of the human need for patience (as well as other virtues) is the imperative for humans to re-align themselves with the patient character of God's creation. This is not to claim that nothing is distinctive about human patience, but that human patience is more about living into our

59. Ibid., 18.

60. It would be interesting to consider how the virtue of patience varies or not among different species, for certainly it is easily observed among animals. That is not, however, the project here.

61. Harned, *Patience*, 159.

humanity as creatures than into our suspected superiority as a species. All human virtues are more about human creatures' relationship with God than about human status vis-à-vis other creatures.

That said, Harned develops a coherent account of the different aspects of patience, which he delineates as disciplined endurance of adversity, undying expectancy of God's Kingdom, unselfish forbearance of the faults of oneself and others, and realistic perseverance in faithful living.[62] My own view differs somewhat, following my different starting position. I see patience as having three fundamental aspects, all of which stem from the character of our existence as creatures and all of which have ecological consequences. Patience is enduring adversity; waiting (and looking) for moments of grace; and, finally, the attentive appreciation of something without desiring its mastery, possession, or improvement.

The first aspect is that of endurance: bearing the ills of life for the sake of faithfulness to God. It may be that God has sent the trial (although Christians should hesitate to move too quickly to this judgment) or that the adversity can only be combated by vicious or unfaithful actions. Thus Christians in the western United States may endure the threats of puma attacks, for instance, because the only certain way of eliminating the threat would be to eliminate the animals, an action clearly in conflict with the command to care for creation.[63] Or Christians in the southeastern part of the country may decide to endure the nuisance of mosquito bites and the slight risk of mosquito-borne disease rather than buy a propane-fueled vacuum that promises to suck up every mosquito in one square acre.[64] This aspect of patience contrasts most sharply with the dictum of late Western capitalism that any desire can and should be satisfied, any risk averted, and the means for the goal should be far greater than necessary.

62. Ibid., 118.

63. Bears nearly always avoid confrontation, and attack only when challenged. The big cats, however, do not seem to distinguish between humans and other food sources, and will sometimes track and stalk people as prey.

64. One model is called the "Skeeter-Vac." Other models have names like warships: "The Liberty" and "The Defender."

It is not the case that Christians must never take action against "natural" threats or hardships, especially in defense of human life. But the importance of patience should proffer two cautions: First, if we are considering measures that harm the non-human creation, we must be as conservative as possible, and remain open to the alternative of enduring rather than easing the hardship. Second, we should be aware that if life is extremely comfortable, our virtue might well be impaired.[65] Sometimes, therefore, we should undertake an ecological reform as a discipline, rather than purely as a means to saving forests or oceans.[66]

The second aspect of Christian patience is quintessentially eschatological. Resting on faith in God's loving providence and promise of the Kingdom of Heaven, patience remains hopefully alert to signs of the Spirit while awaiting the coming of Christ. It is crucial to the ability to continue eco-discipleship in the face of apparent failure. This aspect of patience is the core of Christian witness, for the patient community of faith perseveres in resisting evil and positing signs of God's grace in the midst of a fallen and violent world. Patience of this sort is exemplified in the life of Paul Farmer, an American physician who has spent years providing medical ministry to poor Haitians against impossible odds, and in the work of Floresta, a Christian organization that plants trees in regions where deforestation intensifies the poverty of communities and their land. This patience not only requires hope, but it also *generates* hope; for the persistence of peaceful struggle itself points to God's presence in the midst of violence and sorrow.

---

65. The converse is not necessarily true. Enduring significant hardship does not identify one as virtuous. Yet, that point cannot rightly or easily be raised by people who already sit in the catbird seat. Thus, an affluent white theologian should be extremely hesitant to dispute the emphasis on reversal raised by theologians in the "two-thirds world." See also Yoder's comment on pastoral patience in "Patience as Method," 24–42.

66. This may provide a "rule-of-thumb" when two conflicting practices seem equally harmful to the environment: paper versus plastic grocery bags, for instance. If it is truly impossible to determine which one more effectively witnesses to God's care for creation, it is likely best to pick the option that is least convenient, in order to develop stronger habits and attentiveness to our work. In this case, the least convenient choice is to bring one's own bags for reusing over and over.

Finally, the third aspect of patience is appreciation, an attitude of receptive delight to the particularity of something. It might be expressed as gratitude and doxology (though not romanticism), connected with humility and responsiveness toward God's grace. It is neither the calculating gaze of the colonial gold-hunter, who sees a river only as a route to financial gain, nor the starry-eyed gaze of the Romantic, who yearns for a return to the Garden of Eden. In American history, this sort of patience is notable by its absence; it could have reveled in the grandeur of Niagara Falls and Yosemite, for instance, without needing to "improve" the falls with concrete and hydro-engineering, and to "perfect" Yosemite Valley by removing its Native American residents.[67] It is, rather, an appreciation of the beauty, complexity, and diversity of the biophysical world that includes acknowledgement of nature's fallen character. Wendell Berry and Annie Dillard, both Christians who write in very different ways about land and creatures, display this aspect of patience in their work. *Pilgrim at Tinker Creek*, for instance, describes the skill of watching closely—not waiting for a particular event, but simply to take in the particularity and otherness of what she sees.

The second and third aspects of patience are closely allied with loving attention, described by Iris Murdoch as a "just and loving concentration upon some individual or situation."[68] Attention is a patient way of seeing, acknowledging the otherness of what is seen and appreciating the divine reflection in God's creation. When we "seek and serve Christ in all persons," we pay *attention* to the particular other.[69] We hold ourselves alert to her signals of joy, sorrow, or suffering; we receive her without consuming or enclosing her. Likewise, when we pay close attention to the world we gain an increased sense of the reality of its inhabitants, relationships, and needs. In Harned's words, "we wait upon the world."

---

67. The constructed character of American "nature" is well documented in Wilson, *Culture of Nature*, and Cronon, *Uncommon Ground*.

68. Harned, *Patience*, 127. Iris Murdoch borrows the term "attention" from Simone Weil, *Waiting for God*, 73. My account is quite similar to Harned's here, except that he again limits the arena of patience to human interaction.

69. Episcopal Church, *Book of Common Prayer*, 306.

All three aspects of patience together form the core of Christian witness, for the human virtue of patience reflects (but dimly) the patience of God.[70] And in the arena of ecological discipleship, patience teaches Christians to inhabit the earth, to share its witness to the creator, rather than (try to) dominate it.[71] It enables Christians to persevere in their concrete acts of witness to the redemption of all creation, despite the continued onslaught against animals, species, and ecosystems. We can see this sort of patience in Christina García's novel *The Agüero Sisters*.[72] Ignacio, a Cuban naturalist living in the early twentieth century, is walking along the beach when he encounters a leatherback turtle; he reacts with awe and protectiveness, rather than his customary love-as-control:

> It was then I saw her. Her ridged back and the enormity of her flippers made identification easy, especially in the moonlight. She was over eight feet long, a half ton of slow magnificence. The leatherback turned her wrinkled, spotted neck and gazed at me, as if gauging my trustworthiness. I could see her eyes clearly, the inverse widow's peak of her beak . . . After what seemed interminable digging, the giantess brought her hind flippers together, craned her neck forward, and began to sway slightly to a private rhythm, finally laying her eggs in the sand . . . At dawn, a fat scavenging gull dropped onto the leatherback's buried nest. I cursed the bird and threw a fistful of sand at it. A moment later, more gulls appeared . . . What choice did I have? I sat on the leatherback's nest all that day and all the next night, guarding the eggs from predators, guarding the eggs for her.[73]

Ignacio here displays the patience of endurance (sitting on a sweltering beach for eighteen hours), hopeful waiting, and appreciative attention.

---

70. See Yoder on God's patience. Yoder, *Original Revolution*, 65.

71. Ernst Conradie borrows the term "ecodomy" from Geiko Müller-Fahrenholz to describe this kind of inhabiting. He cites Paul, as well, who uses it for the building up of Christian communities (2 Cor 13:10; Rom 14:19; Eph 2:21). Conradie, "Steward or Sojourners," 162.

72. García, *Agüero Sisters*.

73. Ibid., 94.

Moreover, he does not regard his action as a choice, but a "natural" response to what his patience enabled him to see.

Patience, as with all virtues, is characteristic of communities even more than of individuals. The church, by its very identity, is a patient community: "we await his coming in glory." The church narrates, practices, and teaches patience as a communal virtue; the community and its individual members acquire patience as they are formed by the stories (Scripture) and practices (sacraments as well as outreach and ecological practices) of that same community. Of course patience cannot stand alone; it is part of the constellation of virtues that make up a Christian life. Prudence, or wisdom, determines when patience is appropriate (its appropriateness in an "environmental crisis"), and when it is not. Courage makes patience possible in the face of danger or hostility. Justice ensures that patience is not merely endurance for its own sake, but perseverance toward right relationships with humans and nonhumans alike.

The importance of patience should not be enlisted in favor of any sort of passivism, or of waiting for God to rescue humans from their misdeeds; actions ought not be judged based on short-term effectiveness in the world. If, as Christians, we relinquish our grasp of environmental destruction as a crisis, and understand it as a call to strengthen the faithful virtue of faith communities, we are able to persevere and move forward in the frightening and laborious tasks of changing the way we live. Patience is essential to these tasks. What Wendell Berry writes about terrorism is equally applicable to environmental issues:

> This . . . is the description of an emergency. It is moreover an emergency of the worst kind: one that cannot be resolved by "emergency measures." It is an emergency that calls for patience, and to be patient in an emergency is a hard requirement. But patience is what we must have if we hope to complete our work.[74]

74. Wendell Berry, *Citizenship Papers*, 179.

CHAPTER 5

CONCLUSION

THIS BOOK argues that Christian faithfulness requires ecological disciple-ship on the church's own terms. The point is not to make the church a participant in the "environmental movement," but to make the church more faithful by including the eschatological import of creation in its performance of worship—worship that consists not merely of weekly services, but also consists of a "way" of life that praises and witnesses to Father, Son, and Holy Spirit.

Chapter 1 reviews several factors—political, cultural, and historical—that prevent the churches from more forceful response to ecological issues, including the insufficiency of education, the difficulty of implementing significant resistance to dominant cultural mores, and the cultural weakness of moderate to liberal church leaders. The theo-logical impediment to environmental response, though, is this project's focus. Christians often regard social ethics as directions for contribution toward the management of social problems, or as directions for building the reign of God on earth. These approaches result from the gradual shift in eschatology from understanding the Kingdom of God as something to be prepared for and prayed for, to viewing it as a human project, more or less on God's behalf. Neither approach, however, is capable of generating the faithful perseverance required in caring for God's creation. So, then, the "environmental crisis" is posed as a problem to be solved; its very size and complexity drives Christians into a position of helpless resignation.

The way beyond this impasse is to regard ecological issues as part of Christian witness to God's work rather than exclusively human work—in other words, to view these issues eschatologically. Most works of Christian eco-theology or ecological ethics, however, neglect the role of eschatology in Christian ethics. Three theologians who address eschatology and ecology are Larry Rasmussen, Catherine Keller, and Rosemary Radford Ruether. While their treatments of ecology and eschatology are quite different, they share a certain functionalism with regard to religion: Christian faith is evaluated and reconstructed according to its facility for encouraging earth-friendly action. Moreover, each of them rejects basic elements of Christian eschatology. Rasmussen outlines an eschatology devoid of Jesus Christ; Keller rejects the sovereignty of God, while Ruether rejects any notion of an afterlife. It became clear that there is room for an approach to ecology and eschatology that affirms the central doctrines of Christian eschatology as well as creation.

These doctrines point to the uniqueness of the Christian creation story and its incompatibility with the "common creation story" as advocated by Thomas Berry, Brian Swimme, and others. The second chapter notes that just as there is not a universal, "neutral" understanding of nature, there is also not a universal, "neutral" understanding of why nature came to be, of its purpose, moral status, and so forth. For Christians, the primary focus of questions about creation and eschatology is the Triune God who brought all things into being through Jesus Christ. God's creation is an ongoing act of freedom and graciousness, through which the biophysical universe came into being and is sustained through time. Yet, freedom does not entail caprice; God desires and freely commits to a loving, providential, and sovereign relationship with God's creation, as testified to in Scripture. God's creating is an expression of God's self-giving love that is divinely faithful; God never abandons the world or tires of caring for its inhabitants. Instead, God's creation of the universe is part of the constant divine intention that the world be created, sustained, redeemed in Jesus Christ, and ultimately brought to a consummating communion with God.

The Creator God is Triune: from the Church's beginning, Christians have affirmed the agency or mediation of Christ in the act of creation, and the activity of the Spirit in sustaining and enlivening creation. The trinitarian character of God's creating work points to the integral christological connection between creation and salvation, a connection strongly expressed in Col 1:15–20. Jesus Christ is the one through whom God creates and orders the world, and through whom God will one day glorify the world. Creation is both christological and eschatological. Therefore, creation can only truly be seen *through* the Trinity: it is only through an understanding of God's activity through Christ in the Spirit that creation can be understood. It is also important to acknowledge God's creating *ex nihilo*, because nothing external to God limits or constrains God's work. Creating the universe is *not* as a sculptor shaping clay, or even as a geneticist "producing" hybrid mice. Therefore, both the world we live in *and* the world to come bear a strong element of mystery and surprise. Just as the natural processes and "laws" of this world originate in God's loving will, so does the nature of the eschaton. Affirming creation *ex nihilo* entails that there is no evil that precedes, overcomes, or can outlast the good; *everything* is ultimately under God's reign. It also reminds Christians that time and space, too, are part of the creation that imposes limits on all creaturely existence—though not on God.

The result of God's creating is what we experience as the universe and all its members: species, relationships, ecosystems, soil, emotions, and virtues. This creation is good, although deeply disordered and corrupt. In a dim and fractured way, it reflects its creator and its creator's intentions. It is not a single abstraction called "nature," but an amazing and wonderful diversity of biophysical specificities. It is important both to acknowledge that diversity and to be cautious about ascribing divine intent or significance to particular differences. Too often humans (including Christians) have read difference in terms of inferiority. Christians *can* attest that God's creations differ and converge in ways we do not comprehend; they can attest that differences do not necessitate conflict or inequality, that creation is as much a web of relationships as a pluriformity of creatures, and that all of God's creations are united in their common origin and destiny in Christ.

Nonetheless, creation does not match God's intentions properly; the Fall is a reminder that all of creation is deeply disordered and corrupt. While Christian understandings of the Fall have varied across the centuries, there is strong theological ground for regarding the universe as—in some way—fallen away from God's plan. The prevalence of suffering, violence, and waste in both human and nonhuman arenas clearly contradicts God's desire for the world as expressed in the Old and New Testaments. The importance of acknowledging the Fall is the recognition that evil *is* present in the world, that creation both affirms and denies God at the same time, that humans and nonhumans are never completely distinguishable, and that human activity matters in the world beyond human culture. This means that nothing in the created world, no human power, no evolutionary shift, can "cure" creation. On the other hand, humans are not responsible for all the suffering in creation; were all humans to vanish suddenly, suffering and violence would not also disappear. God's love is precisely directed toward all those who suffer; creation needs redeeming, and God has sent Jesus for this purpose. In itself, therefore, biophysical reality cannot function normatively for Christians. (This cuts both ways, ethically; neither is something good just because it is "natural," nor is something bad just because it is "unnatural.")

Finally, creation is eschatological. Jesus Christ is the redeemer of the world, and not only of human "souls." Biblical testimony as well as ancient and contemporary theologians argue decisively that nonhuman creation serves God, praises God, and will partake in the Kingdom of God. The purpose of creation is to glorify God, and that purpose is fulfilled partially in earthly life, and only completely in the final consummation. So the universe is doxological and eschatological in its very being. Because creation is distinct from God, not an extension of Godself, it receives its existence, inspiration, and purpose as trinitarian gift.

Chapter 3 presents a conception of Christian life as witness. Witness means a particular understanding of discipleship in which the communal lives of the disciples testify, through character, worship, and action, to the Kingdom of God as inaugurated, preached, demonstrated, and promised by Jesus Christ. Employing "witness" as the ethical hermeneutic performs several key functions: it emphasizes the sovereignty of God; it avoids the slide into despair in our ecological lives; it ensures the Christian unity of faith and action; it demonstrates the embodied connection between creation and redemption; and it facilitates a sensitivity to social location. These are crucial advantages for Christian ecological ethics.

Witnessing to the Kingdom requires an informed faith in the promises of God as related in Scripture and the life of the church. This is not knowledge in the scientific sense, nor purely intellectual acceptance of cognitive propositions. As Jon Sobrino writes, praxis—following Jesus Christ in obedience—is the means to grasping the nature of the Kingdom of God. Praxis and understanding form a spiral process: as Christians testify to Christ's redemption of all creation, they begin to understand what that means, and as they grow in understanding, their testimony deepens and strengthens. This process (or journey) occurs in direct confrontation to the forces of the anti-Reign, which depends on violence, fear, and death for their power. Yet, confrontation with the anti-Reign also displays the Reign of God by counter-example: the more that oppression and destruction challenge the disciples, the more the disciples understand the importance of liberation, reconciliation, and peace.

Memory as well as praxis shapes Christian understandings of the eschaton, especially with regard to ecological ideals. Environmentalists need to relinquish their nostalgia for a "green age," typically viewed as the time before oil refineries and toxic-waste dumps came to sully the landscape. Not only was the pre-industrial world full of suffering and waste, but the eschaton is also not simply the fulfillment of human desires for a cleaner environment. Our hope must spring not from romantic wilderness landscapes, but from the resurrection of Jesus Christ who comes again as Lord of all creation. On the other hand, the Kingdom as described in Scripture does not obliterate the existing world, but incorporates and transforms the old creation into the new. The wounds of Jesus do not disappear after the resurrection, nor will the scars on the face of the earth. For Christians, the past, present, and future are brought together in Christ. So Christian understanding of the Kingdom incorporates a clear-eyed view of imperfect earthly life. All of creation comes under Christ's reign—the beauty as well as the ugliness. All of it will be transformed and redeemed in ways we cannot imagine, yet can faintly grasp through the prophets and Jesus himself.

The character of that Kingdom is revealed to include peace, justice/liberation/reconciliation, material abundance, righteousness, and communion with God. Peace is the ultimate reality of our world, because it was created out of the irenic, extravagant love of Father, Son, and Holy Spirit. Peace is part of every biblical prophecy about the Kingdom of God, and this peace obtains across all boundaries—including species boundaries. The peace of Christ is the unity of Christ in the source, savior, and destiny of all creatures. The Kingdom's justice results from liberation of the oppressed and reconciliation of enemies. Again, because nothing is excluded from Christ's redeeming work, justice and reconciliation cannot be limited to human beings. Relationships among all of God's creatures are reconciled into the unity of Christ.

The abundance of the Kingdom is signaled by Old Testament texts, the book of Revelation, and Jesus' feeding miracles; and it is prefigured by the Eucharist, the foretaste of the eschatological feast. This overflowing fruition of earth contradicts the contemporary emphasis on scarcity of goods and the "inevitability" of hunger and poverty. It contradicts, too, the understanding of the natural world as *essentially* violent and competitive. This abundance, while never attainable on earth, is possible in the freedom and love of God. Finally, everything in the Kingdom participates in trinitarian communion. Christ's lordship over all things puts all things under subjection to God, and draws all creation into the loving communion among Father, Son, and Spirit (1 Cor 15:28). Yet, communion is not union, and the ultimate reality of the universe is Trinity, so participation in the life of the Trinity does not eradicate the distinctiveness of God's creatures; the new creation is, in some ways, continuous with the old.

These descriptions of the Kingdom of God are neither exhaustive nor absolute; they arise from the understandings Christians have held about God's promises to and for the created world. And while those understandings include other characteristics as well, those noted in chapter 3 are necessary both to Christian conceptions of the eschaton as well as to the arguments about ecological discipleship.

The Kingdom of God is a realm of peace, justice, reconciliation, and abundance, to be consummated when the created world is graciously redeemed by the Son in God's own time. How, then, do Christians follow the command to proclaim the Kingdom and live as disciples to Jesus Christ? More specifically, how do Christians engage in the self-sacrificial service that exemplifies the truth of the gospel? Witness is an ethic of response to God's faithfulness to the created world and to God's people, despite the fallen and corrupt character of creation. It is, therefore, obedient to God's commands without presuming to control the results of that obedience. We witness to the reality of Christ himself, and to the work he has already done, is presently doing, and will do in redeeming the world.

It is therefore more accurate, at least for persons on the "upperside" of history, to describe Christian life as "witnessing" or "testifying" to the Kingdom rather than "building" the Kingdom or "participating" in the Kingdom's completion. What is at stake in the language issue is the essential and constant awareness of our creaturely dependence upon God, that cautions us against trying to invent the Kingdom according to our own desires, or to force the eschaton onto the world—or both.

Jon Sobrino describes witness as positing signs of the Kingdom in Jesus' wake—for example, comforting those in distress without imagining we can eliminate distress, and struggling for reconciliation and forgiveness without supposing we can eradicate conflict. The coming of the Kingdom requires a conversion, a grasping of the reality that all are called to live in accordance with the eschaton. This means that Christians need not rely on means-end rationality to determine a course of action, but on Christian prudence. Prudence is practical reasoning toward the action that best testifies to God's love for the world, even if the short-term prospect is dismal failure.

Witness is not only prudent, but is also costly, for (in Sobrino's terms) the Reign of God opposes the anti-Reign, and the anti-Reign defends itself with fear, violence, and death. Yet, it remains hopeful, even joyous. Christian hope is not optimism about human nature's eventual triumph, or about evolutionary progress. Instead, Christian hope and joy arise from the resurrection, in which Jesus' love overcame evil and death. Christian discipleship also bears a continuing openness to eschatological possibilities: the possibility that new circumstances might call for new forms of witness, for more creative ways of responding to God's grace, or for further refinement of discipleship practices. Finally, witness is fundamentally communal, because faith is nurtured, expressed, and fulfilled in the community of believers.

An ethic of witness thus described enables the church to move past the management/despair dilemma that has so often enfeebled its responses to ecological issues. This is because witnessing to the Kingdom emphasizes the sovereignty of God over all of creation and the priority of Christ in inaugurating and consummating the eschaton. Yet, witness also treats human actions as highly significant insofar as they establish signs to the eschaton. Furthermore, witness has a strong tradition in Christian history, explicates the unity of faith and action evident in Jesus and the apostles, and makes vivid the connection between creation and redemption. Articulating Christian ethics as witness allows me to show that, contrary to what is expressed by many secular environmentalists, small actions work. They do not work in the sense of effecting short-term ecological improvement, but in the sense of indicating the possibilities already enabled by God.

Chapter 4 addresses the ecclesial nature of witness and the specifics of ecological discipleship. New Testament scholarship makes clear that the addressee of Jesus' commands is the church; Jesus' commands and exhortations are not directed at individuals, but at the community of believers. Moreover, as the addressee of God's mandates and God's promises, the church is the eschatological precursor of the heavenly city. The church, like all of creation, can only be understood eschatologically; it is the earthly sign of the heavenly reality. Its very presence anticipates the fulfillment of God's intentions for all of creation.

Faith is communal, because Christianity sees the human subject itself as communal. What has been called the autonomous individual self is, in fact, the construction of various relationships in a particular historical construct. Humans (as well as nonhumans) are embedded at the deepest level with the lives of other creatures—socially, personally, biologically. After all, part of the implication of creation's reflecting its trinitarian creator is that all of creation participates in the reality of "beings in communion." The being-in-communion of creatures is not independent; creation does not exist at any moment apart from its creator. Instead, the existence and identity of everything is unilaterally dependent on God's freedom and generosity.

The church, then, is the community of worship formed by Christ as God's Son, through the intertwined movements of human events and natural history, which produces disciples through acts of witness enabled by the Spirit. The church is not the "midwife" of the eschaton, but its herald; it does not create the Kingdom, but shows how the Kingdom has already been inaugurated through the self-sacrifice of Jesus Christ. The church neither functions as a vehicle to entice people toward responsible eco-behavior nor as a vehicle to "save" creation (in Christian or non-Christian terms). Again, just as the rest of creation, the church has its creator and purpose extrinsic to itself. As the eucharistic body of Christ, it is created and re-created (or enacted) through the performance of Eucharist. Eucharist, moreover, brings together past, present, and future as earthly and heavenly reality meet in the sacrament. The members of the church are gathered into Christ at the Eucharist just as all of creation will be gathered into Christ at the eschaton.

So the church is repeatedly enacted through Christ's sacrifice for the world, the event in which all things are forgiven and made new; consequently, the church is predicated on God's gift of forgiveness and reconciliation. God's forgiveness means that the church is not bound by its sinfulness, but is enabled through the Spirit to live into new possibilities of peace and abundance despite the hegemony of fear and violence in the world. Living into those possibilities consists in "works" (outreach, ethics, evangelism) and worship, both of which are better viewed as gifts from God than as programs the church develops. Works and worship are deeply interdependent: works are part of the church's doxological response to God, and worship constitutes the witnessing body that is the church. So everything the church does, not just outreach or evangelism, is shaped by its witness to the eschaton; everything must be formed eschatologically. Churches are called in their worship, works, Bible study, committee meetings, fund-raising—in everything—to witness to Christ's lordship over the Kingdom of peace, abundance, justice/liberation/reconciliation, and righteousness for all creation.

As the Kingdom is not limited to human souls, or even human bodies, so ecological witness includes nonhuman creation. Yet, Christian ecological witness differs from secular environmentalism. The church is not recruiting people of good will to save the world (environmentally), but is demonstrating in small, yet significant practices, how a (theologically and environmentally) saved world might operate. Rather than endorsing the common view of life on earth as necessarily conflictual and competitive, therefore, churches can exhibit a more peaceful relationship with their biophysical surroundings. These peaceful relationships can protect local land, animal species, and ecosystems, even though they might offend human neighbors by violating cultural prescriptions about land maintenance. Witnessing to peace also entails nonviolence (or at least reduced violence) toward animals, and justice to human workers, thus promoting vegetarian meals, gentler farming methods, and just treatment of farm workers, and fair-trade products. In particular, the eucharistic offerings of bread and wine should be produced under conditions of just labor and sustainable farming methods. What we offer to God and take into our bodies as the body and blood of Christ should testify to God's care for the earth and Christ's redemption of all creation.

Further, the church can posit signs of the eschatological bounty of earth by willingly sharing its goods, rather than hoarding or overconsuming from fear of scarcity. Ironically, testifying to the Kingdom's abundance thus requires a cheerful self-denial—a reduction in use of fossil fuels, of goods that cannot be recycled, of landfill space, and so forth.

These are only initial steps in "green" discipleship. They indicate, nonetheless, how Christian ecclesial witness differs from most secular environmentalism. The church does not attempt to reform the world by regulation or force, although its individual members may work for better environmental policies. Rather, as an alternative, eschatology polity, the church itself posits signs of the Kingdom of God by living as though God's promises are true. The practices of the church demonstrate that, through Christ, all of creation is loved and redeemed by God.

The practices of the church demonstrate that, through Christ, all of creation is loved and redeemed by God. In contrast, then to the Christian ecologies that affirm a created, but static, universe, is the ecological stance that affirms a created biophysical reality that is not yet completed. In the latter view, God's continuing activity is completed in the return of Jesus Christ at the eschaton. Understanding the creation in which we live, therefore, requires understanding not only its beginnings, but also its consummation in the Kingdom of God. And understanding the Christian vision of the Kingdom in turn requires a standpoint within the church—a standpoint centered in Eucharist, praxis, and communal faith. So, after all, it is not environmentalists alone who can "solve" the "environmental crisis." It is, rather, the church that can undertake the difficult, costly efforts of witnessing to Christ's redemption of the world.

BIBLIOGRAPHY

Achterhuis, Hans. "Scarcity and Sustainability." In *Global Ecology: A New Arena of Political Conflict*, edited by Wolfgang Sachs, 104–16. London: Zed, 1995.

Adams, Carol J. *The Sexual Politics of Meat: A Feminist-Vegetarian Critical Theory*. New York: Continuum, 1990.

Allchin, A. M. *Participation in God: A Forgotten Strand in Anglican Tradition*. Wilton, CT: Morehouse Barlow, 1988.

Athanasius. *Contra Gentes and De Incarnatione*. Translated and edited by Robert W. Thompson. London: Oxford University Press, 1971.

Augustine. *City of God*. Translated by Gerald G. Walsh et al. New York: Doubleday, 1958.

Bader-Saye, Scott. "Aristotle or Abraham: Church, Israel, and the Politics of Election." PhD diss., Duke University, 1997.

Baker-Fletcher, Karen. *Sisters of Dust, Sisters of Spirit: Womanist Wordings on God and Creation*. Minneapolis: Fortress, 1998.

Banuri, Tariq. "The Landscape of Political Conflicts." In *Global Ecology: A New Arena of Political Conflict*, edited by Wolfgang Sachs, 49–67. London: Zed, 1995.

Basil the Great. *On the Holy Spirit*. Translated by David Anderson. Crestwood, NY: St. Vladimir's Seminary Press, 1980.

Bauckham, Richard. "Eschatology." In *Oxford Companion to Christian Thought*, edited by Adrian Hastings, 206–9. Oxford: Oxford University Press, 2000.

———. *God and the Crisis of Freedom: Biblical and Contemporary Perspectives*. Louisville: Westminster John Knox Press, 2002.

———. "Jesus and the Animals II: What Did He Practise?" In *Animals on the Agenda: Questions about Animals for Theology and Ethics*, edited by Andre Linzey and Dorothy Yamamoto, 49–60. Urbana: University of Illinois Press, 1998.

Beiner, Ronald. *What's the Matter with Liberalism?* Berkeley: University of California Press, 1992.

Bell, Daniel. "The Insurrectional Reserve: Latin American Liberationists, Eschatology, and the Catholic Moment." *Communio* 27 (2000) 643–75.

Benhabib, Seyla. *Situating the Self: Gender, Community, and Postmodernism in Contemporary Ethics*. New York: Routledge, 1992.

Berkhof, Hendrikus. *Christ and the Powers*. Translated by John Howard Yoder. Scottdale, PA: Herald, 1962.

Berry, Thomas. *Technology and the Healing of the Earth*. Teilhard Studies 14. Chambersburg, PA: Anima, 1985.

Berry, Wendell. *Citizenship Papers*. Washington, DC: Shoemaker and Hoard, 1993.

———. *What Are People For?* San Francisco: North Point, 1990.

———, and Norman Wirzba, eds. *The Art of the Commonplace: The Agrarian Essays of Wendell Berry*. Washington, DC: Shoemaker and Hoard, 2002.

Betzer, Just. *Babette's Feast. Le Festin De Babette*. Directed by Gabriel Axel. 1987.

Booty, John E. *Reflections on the Theology of Richard Hooker*. Sewanee, TN: Sewanee Mediaeval Colloquium, 1999.

Boucher-Colbert, Marc. "Eating the Body of the Lord: Eucharist and Community-Supported Farming." In *Embracing Earth: Catholic Approaches to Ecology*, edited by Albert J. Lachance and John E. Carroll. Maryknoll, NY: Orbis, 1994.

Bouma-Prediger, Steven. *For the Beauty of the Earth: A Christian Vision of Creation Care*. Grand Rapids: Baker Academic, 2001.

Bratton, Susan Power. *Christianity, Wilderness, and Wildlife: The Original Desert Solitaire*. Scranton, PA: University of Scranton Press, 1993.

Bridger, Francis. "Ecology and Eschatology: A Neglected Dimension." *Tyndale Bulletin* 41 (1990): 290–301.

Brueggemann, Walter. *The Land: Place as Gift, Promise, and Challenge in Biblical Faith*. Overtures to Biblical Theology. Philadelphia: Fortress, 1977.

Burke, William Kevin. "The Wise Use Movement: Right Wing Anti-Environmentalism." *The Public Eye* (June 1993).

Calvin, John. *Calvin's Commentaries*. Translated and edited by John King. Grand Rapids: Eerdmans, 1948.

Carmody, John. "Ecological Wisdom and the Tendency toward a Remythologization of Life." *Concilium* 4:100 (1991) 94–103.

Carter, Stephen. *The Culture of Disbelief: How American Law and Politics Trivialize Religious Devotion*. New York: Basic, 1993.

Cavanaugh, William. *Torture and Eucharist*. Oxford: Blackwell, 1998.

Charlesworth, James H. *The Old Testament Pseudepigrapha*. Garden City, NY: Doubleday, 1985.

Clifford, Richard J. "The Bible and the Environment." In *Preserving the Creation*, edited by Kevin W. Irwin and Edmund D. Pellegrino, 1–26. Washington, DC: Georgetown University Press, 1994.

Coleman, Roger, ed. *Resolutions of the Twelve Lambeth Conferences, 1867–1988*. Toronto, ON: Secretary General of the Anglican Consultative Council, 1992.

Cone, James. *A Black Theology of Liberation*. Twentieth Anniversary Edition. Maryknoll, NY: Orbis, 1990.

———. *The Spirituals and the Blues*. Maryknoll, NY: Orbis, 1992.

Conradie, Ernst. "Stewards or Sojourners in the Household of God?" *Scriptura* 73 (2000) 153–74.

Conyers, A. J. "Living Under Vacant Skies." *Heaven and Hell*. Christian Reflection (2002) 9–17. Waco, TX: Baylor University, 2002.

Copeland, M. Shawn. "Journeying to the Household of God: The Eschatological Implications of Method in the Theology of Letty Mandeville Russell." In *Liberating Eschatology*, edited by Margaret Farley and Serene Jones, 26–46. Louisville: Westminster John Knox, 1999.

Cronon, William, ed. *Uncommon Ground: Rethinking the Human Place in Nature*. New York: Norton, 1995.

Davis, Ellen. "Preserving Virtues: Renewing the Tradition of the Sages." In *Character and Scripture: Moral Formation, Community, and Biblical Interpretation*, edited by Wm. P. Brown, 183–201. Grand Rapids: Eerdmans, 1992.

Edwards, Denis. *Jesus the Wisdom of God*. Maryknoll, NY: Orbis, 1995.

Edwards, Jonathan. "A Dissertation Concerning the End for Which God Created the World." In *Ethical Writings*, edited by Paul Ramsey, 403–536. New Haven: Yale University Press, 1989.

Eichrodt, Walther. "In the Beginning: A Contribution to the Interpretation of the First Word of the Bible." In *Creation in the Old Testament*, edited by Bernard W. Anderson, 65–73. London: SPCK, 1984.

Episcopal Church of the USA. *The Book of Common Prayer and Administration of the Sacraments and Other Rites and Ceremonies of the Church: Together with the Psalms of David: According to the Use of the Episcopal Church*. New York: Church Hymnal Corp., Seabury, 1979.

Evans, James. *We Have Been Believers: An African-American Systematic Theology*. Minneapolis: Fortress, 1992.

Farrow, Douglas. "Eucharist, Eschatology and Ethics." In *The Future as God's Gift: Explorations in Christian Eschatology*, edited by David Fergusson and Marcel Sarot, 199–216. Edinburgh: T. & T. Clark, 2000.

Fern, Richard L. *Nature, God, and Humanity: Envisioning an Ethics of Nature*. New York: Cambridge University Press, 2002.

FitzGerald, Thomas. "The Holy Trinity and Creation." *Ecumenism* 100 (1990) 9–11.

Fowler, Robert Booth. *The Greening of Protestant Thought*. Chapel Hill: University of North Carolina Press, 1995.

Frey, Christopher. "Eschatology and Ethics: Their Relation in Recent Continental Protestantism." In *Eschatology in the Bible and in Jewish and Christian Tradition*, edited by Henning Graf Reventlow, 62–74. Sheffield: Sheffield Academic, 1997.

Friesen, Will. "What Are We Fighting For?" *Direction* 21.2 (1992) 47–53.

Galloway, Allan D. *The Cosmic Christ*. New York: Harper, 1951.

García, Christina. *The Agüero Sisters*. New York: Ballentine, 1998.

Gilkey, Langdon. "God." In *Christian Theology: An Introduction to Its Traditions and Tasks*, edited by Peter C. Hodgson and Robert H. King, 62–87. Philadelphia: Fortress, 1982.

Glionna, John M, "Napa Growers to Build Housing for Harvesters." *Los Angeles Times*, March 19, 2002.

Goizueta, Roberto S. "Why Are You Frightened?: U.S. Hispanic Theology and Late Modernity." In *El Cuerpo de Cristo: The Hispanic Presence in the U.S. Catholic Church*, edited by Peter J. Casarella and Raúl Gómez, 51–62. New York: Crossroad, 1998.

Gowan, Donald E. *Eschatology in the Old Testament*. Philadelphia: Fortress, 1986.

Green, Garrett. "Imagining the Future." In *The Future as God's Gift*, edited by David Fergusson and Marcel Sarot, 75–84. Edinburgh: T. & T. Clark, 2000.

Greenhouse Crisis Foundation and the Eco-Justice Working Group, National Council of Churches of Christ in the USA. *101 Ways to help save the earth with 52 weeks of congregational activities to save the earth*. New York: NCCC, 1994.

Greenway, William. "Animals and the Love of God." *Christian Century* 117:19 (June 21–28, 2000) 680–81.

Gregorios, Paulos Mar. "New Testament Foundations for Understanding the Creation." In *Tending the Garden: Essays on the Gospel and the Earth*, edited by Wesley Granberg-Michaelson, 37–45. Grand Rapids: Eerdmans, 1984.

Gunton, Colin E. *Christ and Creation*. Grand Rapids: Eerdmans, 1993.

———, ed. *The Doctrine of Creation: Essays in Dogmatics, History, and Philosophy*. Edinburgh: T. & T. Clark, 1997.

Haas, Guenther. "The Significance of Eschatology for Christian Ethics." In *Looking into the Future: Evangelical Studies in Eschatology*, edited by David W. Baker, 325–41. Grand Rapids: Baker Academic, 2001.

Hanby, Michael. *Augustine and Modernity*. New York: Routledge, 2003.

Harding, Sandra. *The Science Question in Feminism*. Ithaca, NY: Cornell University Press, 1986.

Hardy, Daniel W. "Creation and Eschatology." In *The Doctrine of Creation*, edited by Colin E. Gunton, 105–34. Edinburgh: T. & T. Clark, 1997.

Hare, John. "The Virtue of Hope." *Heaven and Hell*. Christian Reflection (2002) 18–23. Waco, TX: Baylor University.

Harned, David Baily. *Patience: How We Wait Upon the World*. Cambridge, MA: Cowley, 1997.

Hauerwas, Stanley. *A Better Hope: Resources for a Church Confronting Capitalism, Democracy, and Postmodernity*. Grand Rapids: Brazos, 2000.

———. *With the Grain of the Universe: The Church's Witness and Natural Theology: Being the Gifford Lectures Delivered at the University of St. Andrews in 2001*. Grand Rapids: Brazos, 2002.

———, and Samuel Wells. "Christian Ethics as Informed Prayer." In *The Blackwell Companion to Christian Ethics*, edited by Stanley Hauerwas and Samuel Wells, 3–12. Boston: Blackwell, 2004.

Hays, Richard B. *The Moral Vision of the New Testament: Community, Cross, New Creation: A Contemporary Introduction to New Testament Ethics*. San Francisco: HarperSanFrancisco, 1996.

Hendry, George. *Theology of Nature*. Philadelphia: Westminster, 1980.

Hildyard, Nicholas. "Foxes in Charge of the Chickens." In *Global Ecology: A New Arena of Political Conflict*, edited by Wolfgang Sachs, 22–35. London: Zed, 1995.

Hooker, Richard. *Of the Lawes of Ecclesiastical Polity*. London: MacMillan, 1865.

———. "Sermon on Pride." In *The Folger Library Edition of the Works of Richard Hooker*. Cambridge: Harvard University Press, 1977.

Hopkins, Dwight. *Heart and Head: Black Theology Past, Present, and Future*. New York: Palgrave, 2002.

Hütter, Reinhard L. "*Creatio ex nihilo:* Promise of the Gift." *Currents in Theology and Mission* 19.2 (1992) 89–97.

———. "The Church: Midwife of History of Witness of the Eschaton?" *Journal of Religious Ethics* 18 (1990) 27–54.

Irenaeus. *St. Irenaeus Against the Heresies*. Translated and edited by Dominic J. Unger. New York: Paulist, 1992.

Isasi-Díaz, María. "Mujerista Theology's Method: A Liberative Praxis, a Way of Life." In *Mestizo Christianity: Theology from the Latino Perspective*, edited by Arturo J. Bañuelas, 175–91. Maryknoll, NY: Orbis, 1995.

Johnson, Elizabeth A. *God For Us: The Mystery of God in Feminist Theological Discourse*. New York: Crossroad, 1994.

Kant, Immanuel. *The End of All Things*. Translated and edited by Lewis Beck White. Indianapolis: Bobbs-Merrill, 1963.

Kärkkäinen, Veli-Matti. *An Introduction to Ecclesiology: Ecumenical, Historical, and Global Perspectives*. Downers Grove, IL: InterVarsity, 2002.

Kearns, Laurel. "What Does Justice Taste Like? The Churches and Fair-Trade Coffee as Eco-Justice Praxis." Annual meeting of the American Academy of Religion, San Antonio, TX, November 1994.

Keller, Catherine. *Apocalypse Now and Then: A Feminist Guide to the End of the World*. Boston: Beacon, 1996.

———. "Composting our Connections: Toward a Spirituality of Relation." In *The Greening of Faith: God, the Environment, and the Good Life*, edited by John E. Carroll, 165–73. Hanover, NH: University Press of New England, 1997.

———. "Power Lines." In *Power, Powerlessness, and the Divine*, edited by Cynthia L. Rigby, 55–77. Atlanta: Scholars, 1997.

Keller, Evelyn Fox. *A Feeling for the Organism*. New York: Freeman, 1983.

LaCugna, Catherine. *God for Us: The Trinity and Christian Life*. New York: Harper Collins, 1991.

Lewis, C. S. *English Literature in the Sixteenth Century: Excluding Drama*. Oxford: Oxford University Press, 1954.

Libânio, João Batista. "Hope, Utopia, Resurrection." In *Systematic Theology: Perspectives from Liberation Theology*, edited by Jon Sobrino and Ignacio Ellacuría, 279–90. Maryknoll, NY: Orbis, 1996.

Linden, Eugene. *The Parrot's Lament: And Other True Tales of Animal Intrigue, Intelligence, and Ingenuity.* New York: Dutton, 1999.

Lindsey, Hal. *The Late, Great Planet Earth.* Grand Rapids: Zondervan, 1970.

————. *The Road to Holocaust.* New York: Bantam, 1990.

Linzey, Andrew. *Animal Theology.* Urbana: University of Illinois Press, 1995.

Lohmann, Larry. "Resisting Green Globalism." In *Global Ecology: A New Arena of Political Conflict,* edited by Wolfgang Sachs, 157–69. London: Zed, 1995.

Louth, Andrew, ed. *Ancient Christian Interpretations of Genesis.* Chicago: Fitzroy Dearborn, 2001.

Lowes, AnthoNew York. "Up Close and Personal: In the End, Matter Matters." In *Earth Revealing, Earth Healing: Ecology and Christian Theology,* edited by Denis Edwards, 125–44. Collegeville, MN: Liturgical, 2001.

MacIntyre, Alasdair. *Dependent Rational Animals: Why Human Beings Need the Virtues.* Chicago: Open Court, 1999.

Maddex, Jack P. Jr. "Proslavery Millennialism: Social Eschatology in Antebellum Southern Calvinism." *American Quarterly* 31 (1979) 46–62.

McCabe, Herbert. *What is Ethics All About? A Re-evaluation of Law, Love, and Language.* Washington, DC: Corpus, 1969.

McDaniel, Jay. "'Where is the Holy Spirit ANew Yorkway?' Response to a Skeptical Environmentalist." *Ecumenical Review* 42 (1990) 165.

McDonaugh, Sean. *Passion for the Earth.* Maryknoll, NY: Orbis, 1994.

McFague, Sallie. *The Body of God: An Ecological Theology.* Minneapolis: Fortress, 1993.

McNeil, Genna Rae. "Waymaking and Dimensions of Responsibility." In *The Courage to Hope: From Black Suffering to Human Redemption,* edited by Quentin Hosford Dixie, 63–73. Boston: Beacon, 1999.

Mead, G. H. *Mind, Self, and Society from the Standpoint of a Social Behaviorist.* Chicago: University of Chicago Press, 1934.

Merchant, Carolyn. *The Death of Nature: Women, Ecology, and the Scientific Revolution.* San Francisco: Harper & Row, 1980.

Moltmann, Jürgen. *Theology of Hope: On the Ground and the Implications of a Christian Eschatology.* Translated by James W. Leitch. New York: Harper & Row, 1967.

————. *The Trinity and the Kingdom.* Translated by Margaret Kohl. SanFrancisco: Harper & Row, 1981.

Montefiore, Hugh, ed. *Man and Nature.* London: Collins, 1975.

Muddiman, John. "A New Testament Doctrine of Creation?" In *Animals on the Agenda,* edited by Andrew Linzey and Dorothy Yamamoto, 25–32. Urbana: University of Illinois Press, 1998.

Nash, James A. *Loving Nature: Ecological Integrity and Christian Responsibility.* Nashville, TN: Abingdon, 1991.

Niebuhr, H. Richard. *The Responsible Self.* New York: Harper & Row, 1963.

Northcott, Michael. "Do Dolphins Carry the Cross? Biological Moral Realism and Theological Ethics." *New Blackfriars* 84 (2003) 540–53.

——. *The Environment and Christian Ethics.* New York: Cambridge University Press, 1996.

Ochs, Peter. "Genesis 1–2: Creation as Evolution." *The Living Pulpit* (April–June 2000) 8–10.

Oelschlaeger, Max. *Postmodern Environmental Ethics.* Albany, NY: State University of New York Press, 1995.

O'Loughlin, Thomas. "Ecotheology and Eschatology." *Ecotheology* 9 (1999) 71–80.

Phan, Peter. "Woman and the Last Things." In *In the Embrace of God: Feminist Approaches to Theological Anthropology,* edited by Ann O'Hara Graff, 206–28. Maryknoll, NY: Orbis, 1995.

Pinches, Charles. "Eco-Minded: Faith and Action." *Christian Century* 115.22 (1998) 755–57.

Polkinghorne, John. *The God of Hope and the End of the World.* New Haven: Yale University Press, 2003.

Ponting, Clive. *A Green History of the World: The Environment and the Collapse of Great Civilizations.* New York: St. Martin's, 1992.

Rasmussen, Larry. "Adam, Where are You?" *The Clergy Journal* 76.2 (1999) 6.

——. "Chase's Sabbath." *Living Pulpit* 7.2 (1998) 20–21.

——. *Earth Community, Earth Ethics.* Maryknoll, NY: Orbis, 1996.

——. "Global Eco-Justice: The Church's Mission in Global Society." In *Christianity and Ecology: Seeking the Well-Being of Earth and Humans,* edited by Dieter T. Hessel and Rosemary Radford Ruether, 515–30. Cambridge: Harvard University Press, 2000.

——. "Hope and the Environment." *Living Pulpit* 1.1 (1992) 19–20.

——. "The Integrity of Creation: What can it Mean for Christian Ethics?" *Annual of the Society of Christian Ethics* (1995) 161–75.

——. *Moral Fragments and Moral Community: A Proposal for Church in Society.* Minneapolis: Fortress, 1993.

——. "Redemption: An Affair of the Earth." *The Living Pulpit* (April–June 1999) 10–11.

——. "Toward an Earth Charter." *Christian Century* 108.30 (1991) 964–67.

Rauschenbusch, Walter. *Christianizing the Social Order.* New York: Macmillan, 1921.

Roberts, James, and Alexander Donaldson, eds. *The Ante-Nicene Fathers: Translations of the Writings of the Fathers down to A.D. 325.* Vol. 1. Grand Rapids: Eerdmans, 1978–1981.

Rossing, Barbara. "River of Life in God's New Jerusalem: an Eschatological Vision for the Earth's Future." In *Christianity and Ecology: Seeking the Well-Being of Earth and Humans,* 205–24. Cambridge: Harvard University Press, 2000.

Ruether, Rosemary Radford. *Gaia and God: An Ecofeminist Theology of Earth Healing.* San Francisco: HarperSanFrancisco, 1992.

——. *Sexism and God-Talk: Toward a Feminist Theology.* Boston: Beacon, 1983.

Russell, David. *The "New Heavens and New Earth": Hope for the Creation in Jewish Apocalyptic and the New Testament.* Studies in Biblical Apocalyptic Literature 1. Philadelphia: Visionary, 1996.

Sachs, Wolfgang. "Global Ecology and the Shadow of 'Development.'" In *Global Ecology: A New Arena of Political Conflict,* edited by Wolfgang Sachs, 3–21. London: Zed, 1995.

Sakenfield, Katherine Doob. "Ruth 4: An Image of Eschatological Hope; Journeying with a Text." In *Liberating Eschatology: Essays in Honor of Letty M. Russell,* edited by Serene Jones, 55–67. Louisville: Westminster John Knox, 1999.

Sanchez, Rene. "In California's Vineyards, Grapes of Wealth and Wrath." *Washington Post,* December 6, 1998, final edition, A03.

Santmire, Paul. *The Travail of Nature: The Ambiguous Ecological Promise of Christian Theology.* Minneapolis: Fortress, 1985.

Sauter, Gerhard. *What Dare We Hope? Reconsidering Eschatology.* Harrisburg, PA: Trinity, 1999.

Schwöbel, Christof. "God, Creation, and the Christian Community: The Dogmatic Basis of a Christian Ethic of Createdness." In *The Doctrine of Creation,* edited by Colin E. Gunton, 149–76. Edinburgh: T. & T. Clark, 1997.

Scott, Peter. "Nature in Christian Theology: Politics, Context and Concepts." In *A Political Theology of Nature,* 3–29. New York: Cambridge University Press, 2003.

Sittler, Joseph. "Called to Unity." *The Ecumenical Review* 14 (1961–1962) 177–87.

———. *Evocations of Grace: Writings on Ecology, Theology, and Ethics.* Edited by Steven Bouma-Prediger and Peter Bakken. Grand Rapids: Eerdmans, 2000.

———. "The Presence and Acts of the Triune God in Creation and History." In *The Gospel and Human Destiny,* edited by Vilmos Vajta, 90–136. Minneapolis: Augsburg, 1971.

Sobrino, Jon. "Central Position of the Reign of God in Liberation Theology." In *Systemic Theology: Perspectives from Liberation Theology,* edited by Jon Sobrino and Ignacio Ellacuría, 38–74. Maryknoll, NY: Orbis, 1996.

———. *Christ the Liberator: A View from the Victims.* Maryknoll, NY: Orbis, 2001.

———. *Jesus the Liberator: A Historical-Theological Reading of Jesus of Nazareth.* Maryknoll, NY: Orbis, 1993.

———. "Spirituality and the Following of Jesus." In *Systematic Theology: Perspectives from Liberation Theology,* edited by Jon Sobrino and Ignacio Ellacuría, 236–56. Maryknoll, NY: Orbis, 1993.

Soper, Kate. *What is Nature?: Culture, Politics and the Non-Human.* Oxford: Blackwell, 1995.

Spivak, Gayatri. "Can the Subaltern Speak?" In *Colonial Discourse and Post-Colonial Theory: A Reader,* edited by Patrick Williams and Laura Chrisman, 66–111. New York: Columbia University Press, 1994.

Standing Liturgical Commission. *Enriching Our Worship: Supplemental Liturgical Materials prepared by the Standing Liturgical Commission.* New York: Church, 1997.

Stassen, Glen H. and David P. Gushee. *Kingdom Ethics: Following Jesus in Contemporary Context*. Downers Grove, IL: InterVarsity, 2003.

Swimme, Brian. *The Universe is a Green Dragon: A Cosmic Creation Story*. Santa Fe, NM: Bear, 1984.

———, and Thomas Berry. *The Universe Story: From the Primordial Flaring Forth to the Ecozoic Era—A Celebration of the Unfolding of the Cosmos*. San Francisco: Harper, 1994.

Taylor, Andrew. "'Gerrard Winstanley's Theology of Nature: the Telos of Creation,' A Common Treasury for All." *ARC* 18 (1990) 101–14.

Thistlethwaite, Susan Brooks. *Sex, Race, and God*. London: Chapman, 1990.

Thomas, Keith. *Man and the Natural World: Changing Attitudes in England* 1500–1800. New York: Oxford University Press, 1996.

Thomashow, Mitchell. "Toward a Cosmopolitan Bioregionalism." In *Bioregionalism*, edited by Michael Vincent McGinnis, 121–32. New York: Routledge, 1999.

Van Houtan, K. S. and S. L. Pimm. "The Various Christian Ethics of Species Conservation." In *Religion and the New Ecology*, edited by D. Lodge and C. Hamlin, 116–48. Notre Dame, IN: University of Notre Dame Press, 2006.

Verhey, Allen. *The Great Reversal: Ethics and the New Testament*. Grand Rapids: Eerdmans, 1984.

Villafañe, Eldín. "An Evangelical Call to a Social Spirituality." In *Mestizo Christianity: Theology from the Latino Perspective*, edited by Arturo J. Bañuelos, 209–23. Maryknoll, NY: Orbis, 1995.

Volf, Miroslav. Review of *Apocalypse Now and Then* by Catherine Keller. *Modern Theology* (1998) 563–65.

Wallis, Jim. *Agenda for Biblical People: A New Focus for Developing a Life-style of Leadership*. New York: Harper, 1976.

Webb, Stephen H. "Ecology vs. the Peaceable Kingdom: Toward a Better Theology of Nature." *Soundings* 79 (1996) 239–52.

White, Lynn. "The Historical Roots of our Ecological Crisis." *Science* 155 (1967) 1203–7.

Wilkinson, Loren. "The New Story of Creation: A Trinitarian Perspective." *Crux* 30:4 (1994) 26–36.

Williams, Raymond. *Problems with Materialism and Culture*. London: Verso, 1980.

Williams, Rowan. "Hooker: Philosopher, Anglican, Contemporary." In *Richard Hooker and the Construction of Christian Community*, edited by Arthur Stephen McGrade, 369–83. Tempe, AZ: Medieval & Renaissance Texts & Studies, 1997.

———. *On Christian Theology*. Oxford: Blackwell, 2000.

———. *A Ray of Darkness*. Boston: Cowley, 1995.

———. "Sharing God's Planet: A Christian Vision for a Sustainable Planet." London: Church House, 2005.

Wilson, Alexander. *The Culture of Nature: North American Landscape from Disney to the Exxon Valdez*. Cambridge, MA: Blackwell, 1992.

Wogaman, Philip. *A Christian Method of Moral Judgment*. London: SCM, 1976.

Wyatt, Peter. *Jesus Christ and Creation in the Theology of John Calvin*. Princeton Theological Monograph Series 42. Allison Park, PA: Pickwick, 1996.

Wynne-Tyson, Jon, ed. *The Extended Circle: A Commonplace Book of Animal Rights*. New York: Paragon, 1989.

Yeago, David. "Messiah's People: The Culture of the Church in the Midst of the Nations." *Pro Ecclesia* 6.1 (1997) 146–71.

Yoder, John Howard. "The Biblical Roots of Liberation Theology." *Grail* 1 (1985) 55–74.

———. *He Came Preaching Peace*. Scottdale, PA: Herald, 1985.

———. *The Original Revolution: Essays on Christian Pacifism*. Scottdale, PA: Herald, 2003.

———. "'Patience' as Method in Moral Reasoning." In *The Wisdom of the Cross: Essays in Honor of John Howard Yoder*, edited by Stanley Hauerwas et al., 24–42. Grand Rapids: Eerdmans, 1999.

———. *The Politics of Jesus: Behold the Man! Our Victorious Lamb*. Grand Rapids: Eerdmans, 1972.

Zizioulas, John D. "Preserving God's Creation: Three Lectures on Theology and Ecology." *King's Theological Review* 12 (1989) 41–45.